Mathematics for Social Justice

Resources for the College Classroom

AMS/MAA | CLASSROOM RESOURCE MATERIALS

VOL **60**

Mathematics for Social Justice

Resources for the College Classroom

Gizem Karaali
Lily S. Khadjavi

Providence, Rhode Island

2010 *Mathematics Subject Classification.* Primary 00-XX.

For additional information and updates on this book, visit
www.ams.org/bookpages/clrm-60

Library of Congress Cataloging-in-Publication Data
Names: Karaali, Gizem, 1974- editor. | Khadjavi, Lily S., 1969- editor.
Title: Mathematics for social justice : resources for the college classroom / Gizem Karaali, Lily S. Khadjavi, editors.
Description: Providence, Rhode Island : MAA Press, an imprint of the American Mathematical Society, [2019] | Series: Classroom resource materials ; volume 60 | Includes bibliographical references.
Identifiers: LCCN 2019000897 | ISBN 9781470449261 (alk. paper)
Subjects: LCSH: Mathematics–Social aspects. | AMS: General. msc
Classification: LCC QA10.7 .M385 2019 | DDC 303.48/3–dc23
LC record available at https://lccn.loc.gov/2019000897

Contents

Preface

Several years ago, Dr. Eva Curry of Acadia University invited us to participate in a panel titled "Proving Hardy Wrong: Research Mathematics with Social Justice Applications". We were tasked with countering G. H. Hardy's characterization of worthwhile mathematics as a purely abstract – and perhaps esoteric – discipline, and challenging his claim that "[a] science is said to be useful if its development tends to accentuate the existing inequalities in the distribution of wealth, or more directly promotes the destruction of human life". Although in his *A Mathematician's Apology* Hardy notes that this quote was in part a rhetorical flourish, it does invite the search for examples of mathematics which serve to counter inequality and improve human welfare.

During the panel, held at the 2011 Joint Mathematics Meetings in New Orleans, we shared each of our own journeys connecting mathematics to social issues, one focused on public school districting and the other on policing practices in Los Angeles. We were deeply moved by the enthusiasm of the audience members and their willingness to seek out ways in which mathematicians could address social justice issues in their own work.

There are, of course, numerous strands to the broad theme "Mathematics for Social Justice", from employing equitable teaching practices to challenging stereotypes and misconceptions about who gets to do mathematics. Hardy's claim requires us, however, to examine course content in particular, as well as the ways that students can be empowered to use mathematical reasoning and tools in social and political contexts.

We thank Dr. Curry for inviting us to that 2011 panel; she started us on a joint path which led to these volumes. In the intervening years, we have been motivated to bring these topics and others into the mathematics classroom. We were delighted to find that many other mathematicians were excited by similar ideas. This volume, together with the companion text *Mathematics for Social Justice: Focusing on Quantitative Reasoning and Statistics*, represents the work of over forty such mathematicians and academics, thoughtful instructors who have developed original course materials with the understanding that today's college students deserve to have the mathematical tools to tackle a wider range of real-world problems.

We have been greatly inspired by our contributors, who have taken on issues from voting rights to racial inequality. We hope that readers will be similarly inspired.

– Gizem Karaali and Lily S. Khadjavi

Part A

Getting Started

1

An Invitation to Mathematics for Social Justice

Gizem Karaali and
Lily S. Khadjavi

Since the 2001 publication of Bob Moses' *Radical Equations: Math Literacy and Civil Rights* [**8**], North American educators have increasingly recognized the significant connection between mathematics education and social justice. Another 2001 publication, *Mathematics and Democracy: The Case for Quantitative Literacy* [**11**], provided the impetus for mathematicians to tackle the visible and invisible roles of mathematics education in civic life and democratic societies. Today scholars from many corners of the academy are working to ensure that students from all backgrounds receive the best mathematics education they can. One way to achieve this goal is to offer our students a range of opportunities for authentic engagement with mathematics addressing issues that are relevant and timely.

As college mathematics instructors, we have the opportunity and the responsibility to help our students develop strong analytical skills and empower them to tackle complex issues. In the K-12 mathematics education literature, the theme of teaching mathematics for social justice has been present for well over a decade now. Starting with Eric Gutstein and Bob Peterson's *Rethinking Mathematics: Teaching Social Justice by the Numbers* [**5**], those school teachers keen on thinking about such issues have a number of pedagogical resources available to them. Among these is an excellent website which provides many sample lesson plans [**10**], Gutstein's thought-provoking *Reading and Writing the World with Mathematics: Toward a Pedagogy for Social Justice* [**4**], and a most readable collection of essays titled *Teaching Mathematics for Social Justice: Conversations with Educators* [13]. All these references would be of interest to a college mathematics instructor, but as of yet, there are few dedicated resources for the college classroom. This book, together with the second volume, *Mathematics for Social Justice: Focusing on Quantitative Reasoning and Statistics*, aims to fill that gap. Indeed we hope to simply initiate this project and expect that many other resources will follow, much as a few of our own contributors were inspired by their work with Science Education for New Civic Engagements and Responsibilities (SENCER) and the Engaging Mathematics initiative.[1]

These two volumes aim to support college instructors who wish to incorporate ideas and instances of various social justice issues into their classroom by providing them with concrete examples of mathematics connected to a range of contexts. An eclectic collection of modules are accompanied by a handful of thoughtful essays on goals, methods, and possible implementation challenges associated with the idea of incorporating social justice themes into the college mathematics classroom. The primary focus of most contributions in this book is on topics where students can ask their own questions, find and analyze real data, use mathematical tools by themselves, and thus draw their own conclusions.

We have communicated with several mathematicians who have thought seriously about incorporating social justice themes into their own teaching and who had developed and used some materials for this purpose. To reach beyond those to whom we directly extended an invitation to submit, we sent out an open call for contributions and received several contributions from many others who have been thinking about similar issues. The result is an extensive, wide-ranging collection of interesting and useable materials suitable for a broad variety of college mathematics courses, touching on such compelling subjects as human rights and racial injustice.

This first volume consists of two sections:

(1) Essays: Perspectives on incorporating social justice themes into the mathematics classroom, written by college mathematics faculty experienced in this topic;

(2) Modules: Detailed and well-developed course materials ready to be adapted to a variety of courses, from precalculus to differential equations, graph theory, and beyond, written by college mathematics faculty who have often included notes about their experience in their own institutional contexts.

The second volume contains modules that fit most naturally into an introductory statistics curriculum or are specifically designed for general education / quantitative reasoning courses.

The modules typically incorporate a series of in-class activities, research assignments, problem sets, or other methods of engaging the students with the relevant mathematics involved. Most are self-contained, with background necessary to understand the context of the issue and suggestions for relevant data sources provided, so that materials can be readily included into any class where the mathematics necessary is being covered.

---

[1]See Engaging Mathematics at `http://engagingmathematics.ipower.com` for a stimulating collection of materials, developed with funding from the NSF, to engage students in civic issues. Most are on sustainability, for use in courses such as statistics or algebra, but others include projects related to racial justice, voting theory, and more, and many involve data collected from local communities.

More specifically, found in each module are:

The mathematical content of the module: What courses would this module work for? What are the prerequisite mathematical ideas / constructs / procedures students should know already? What will be the mathematical value of the module; in other words, what math is being taught / scaffolded / applied / strengthened?

Context / background for the module: underlying social / political / economic context, background information that students or the instructor should know.

Instructor preparation: direct instructions and additional suggestions for the instructor to be able to use the module in the classroom, possible adaptations.

The module itself: the mathematical and social situation as it is to be introduced to the class.

Additional thoughts: further investigation suggestions, how things went for the author(s) when they implemented the module.

Most modules also have appendices containing tables and datasets as well as sample handouts that instructors can use in their own classes.

Between the two of us, we have several years of professional work on the themes of social justice and mathematics, including various panels, presentations, and workshops at the Joint Mathematics Meetings, law and sociology conferences, and elsewhere, along with a handful of related papers and projects. While writing papers, reviewing journal submissions or editing journals, and making professional connections within and outside of mathematics, we had a growing desire to highlight and draw attention to the compelling and innovative work that so many are doing to bring these themes into their own classrooms. After a couple years of dreaming and planning, we are now delighted to bring this volume to fruition.

What exactly do we mean by social justice?

Before we move any further, we should probably define what we mean by "social justice" or, more specifically, by "teaching mathematics for social justice". Justice Potter Stewart's test ("I know it when I see it") will probably not suffice for an audience of mathematicians.

Following the spirit of the International Council of Scientific Unions, we use, here and elsewhere, the term "teaching mathematics for social justice" to encompass all mathematics instruction which aims for improved human well-being. Human well-being in turn is defined as a "context- and situation-dependent state, comprising basic material for a good life, freedom and choice, health and bodily well-being, good social relations, security, peace of mind, and spiritual experience" [7].

As such "social justice" forms a cornerstone of what makes a society good; individuals and communities in such a society are all guaranteed to have certain fundamental rights. Some may offer John Rawls' theory of justice as a possible way to dig deeper into the concept. For others Rawls will not suffice; those will offer other thinkers as dependable guides. For our purposes, however, the essence is captured by the United Nations: "Social justice may be broadly understood as the fair and compassionate distribution of the fruits of economic growth" [9]. This, coupled with Thomas Jefferson's incomplete list of the unalienable rights, "Life, Liberty, and the pursuit of Happiness", is a basis for our view.[2]

Teaching mathematics for social justice is an expansive project. It involves, among other things, what we do in the classroom with our students to create classroom environments where social justice is practiced. It involves exploring the foundational assumptions of our profession, as well as the historical narratives supporting or disrupting today's academic hierarchies. It also involves understanding and dismantling various structural and systemic barriers confronting large groups of students through mathematics. Recent resources such as *Inventing the Mathematician: Gender, Race, and Our Cultural Understanding of Mathematics* [6], *Rehumanizing Mathematics for Black, Indigenous, and Latinx Students* [3], and *Reflecting the World: A Guide to Incorporating Equity in Mathematics Teacher Education* [2], as well as the many thoughtful posts in the American Mathematical Society's *inclusion/exclusion* blog (available at `https://blogs.ams.org/inclusionexclusion/`), may guide mathematics instructors in such pursuits.

The purpose of this book is much more modest. In this volume and the next, we are almost exclusively concerned with the content of college mathematics courses. We know that this is not the only way to teach mathematics for social justice. However, we believe that incorporating social issues into one's curriculum is a low-barrier entry point into the practice.

What is in the book?

When we began work on this project, we envisioned a single book, though packed full of ideas and resources. As we reached out to the mathematics community, we found that there was much interest in and expertise with regard to this approach. The wealth of materials we gathered together could not fit into a single volume. Therefore we decided to split this collection into two volumes. Below we describe the contents of this book. The contents of the second are described in detail in the introduction to that volume.

The essays. This book starts with five reflective essays written by experienced and thoughtful instructors. These essays will encourage and support instructors in the classroom, as well as in discussions with colleagues and administrators. As such they are an integral part of this volume.

Kira Hamman in "Mathematics in Service to Democracy" shares the story of how she came to teach mathematics for social justice. Her personal story is a good place to start for instructors who may not yet be sure why they should explore social justice themes in their classrooms or how they could convince colleagues, students, and administrators that this is a worthy endeavor. As a veteran of the ongoing conversations around quantitative literacy, Hamman brings to this volume a strong and well-supported conviction that mathematicians need to take more responsibility in the education of the next generation, and that teaching mathematics for social justice is a concrete way to do so.

Suppose you are a mathematician convinced that it is a good idea to try to bring social justice issues into your classroom; after all, you are already reading the introduction

---

[2]If we were to follow this line of thought, we should like to incorporate into a comprehensive definition the individual-directed as well as the communal-directed rights of all people. Political philosopher Danielle Allen argues that these are also accounted for in the American Declaration of Independence; see [1].

to a book just about that! Still you might not feel equipped to jump in. Most mathematics instructors have not had much experience leading in-class discussions, especially those where there may not be clear and objective answers. You might wonder about the pitfalls and the myriad ways things can go wrong once you take the plunge. In her essay titled "Preparing for Student Resistance: Rules of Engagement for Sensitive Topics", Lisa Marano explores just these kinds of issues. In particular, we read about the many ways students may resist this approach and a handful of other unfamiliar teaching problems that mathematics instructors may now need to consider. Marano pulls examples and perspective from a course which she developed and has taught for many years. From her experience the reader will gain confidence that the challenges of teaching mathematics for social justice are surmountable.[3]

For many years there has been a lively effort within the mathematics community to incorporate into standard curricula mathematical problems that involve environmental issues. In the essay "Social Justice and Sustainability: Two Perspectives on the Same System", Jason Hamilton and Thomas J. Pfaff argue that that movement should not be independent from the math-and-social-justice movement. More broadly, they argue that the problems facing the current and the future generations on this planet are complex and multifaceted, and they require sustained and nuanced thinking that can clearly make the connections between environmental and social justice issues. We agree. To this end these two volumes contain a handful of modules that revolve around the issues of environmental justice. See in particular the contributions of Archey; Cohen and Pivarski; Henderson and Kose; and Zobitz, Bibelnieks, and Lester in this volume, and Galanthay and Pfaff, and Piatek-Jimenez in the companion text *Mathematics for Social Justice: Focusing on Quantitative Reasoning and Statistics*.

This volume also contains an essay from Victor Piercey who describes a theoretical construct he labels "quantitative ethics". The essay, also titled "Quantitative Ethics", is an exploration into this construct. According to Piercey, quantitative ethics is inquiry into the ethical implications of generating, using, and disseminating quantitative data. We are not only interested in understanding others' data, Piercey tells his students. We also want to make sure that when we ourselves create data or analyze data and present our numbers to others, we are cognizant of the ethical implications of what we are doing. Piercey offers some provocative examples, and there is much to be pondered upon here.

The last essay, titled "Math for Social Justice: A Last Math Class for Responsible Citizens", and written by Dave Kung, is about a semester-long course Kung has developed and taught for many years that explicitly explores and builds connections between mathematics and social justice. Most of the modules we collected together in this volume are designed to fit into already extant and relatively standard course syllabi. Kung has chosen a different path and developed a whole course. In his essay, he describes his reasons for creating this course and shares some of his experiences with teaching it. Instructors who are intrigued by the challenge will appreciate the contributions of

---

[3]Module contributors to this volume also put in significant effort into providing support for instructors who might be interested in using their classroom materials on various issues that may come up. To this end, for instance, several modules have sections on specifics of grading this kind of work; others provide some hints on how best to assign and help support students through projects that have components that might be unfamiliar to a typical mathematician, such as writing prompts and poster presentations. Furthermore each module contains a section on the social and political background of the issue under exploration for the instructors who might otherwise feel unprepared about diving into the topic as non-experts.

Beier in this volume and of Franco and Piercey in the companion book *Mathematics for Social Justice: Focusing on Quantitative Reasoning and Statistics*; each of these modules describes a semester-long course built around social justice.

The modules in this book. The bulk of this volume consists of fourteen modules, exploring a wide variety of social justice themes and offering the opportunity to include them in courses ranging from college algebra to discrete mathematics. Below we give a brief description of each one, to give the reader a feel for the topics and the mathematics involved; tables provided in the next section will allow the interested reader to track down particular subjects of interest.

Dawn Archey in her contribution titled "Sea Level Change and Function Composition" describes a module intended for a college algebra or precalculus course. Basic mathematical ideas such as function composition and similar triangles come in handy as students work to model the effects of global climate change on small island nations. Numerous supporting references regarding climate change are provided, including those for predictions of change in sea level. The main part of the module can be completed in class; there is also a follow-up writing activity which can be assigned as homework. Archey provides a student handout as well as extensive data to support the instructor.

In "Exploring the Problem of Human Trafficking", Julie Beier has designed a module centered around graph theory and the issue of human trafficking. The module makes use of simple constructions from graph theory to allow students to engage with sophisticated problems. Spanning over eleven class periods, it can be modified to fit shorter time limits and may be used with students with any mathematical background, as the prerequisites, both for the social context and the mathematics involved, are built into the project described. Students gain a deeper understanding of the problem of human trafficking, are introduced to network theory, and develop an appreciation for how mathematics can help us tackle challenges in today's world.

Geoffrey Buhl and Sean Kelly present students with the challenge of creating voter districts which are deemed fair in their contribution titled "Evaluating Fairness in Electoral Districting". The issue of gerrymandering provides the political context for this module, originally designed for a general education course but also adopted in an environmental science workshop. With hands-on materials, students create districts and then mathematically evaluate the fairness of their construction, given geometric compactness measures and an index for measuring voter power. Buhl and Kelly provide background for these types of measures, as well as structured handouts and discussion questions. Given the contributions of mathematicians such as Moon Duchin to related conversations (see [9] for recent coverage of her work), this module may be a great segue to other mathematical topics as well as more engaged discussions of mathematics and its role in the public sphere.

In "Modeling the 2010 Gulf of Mexico Oil Spill", Steve Cohen and Melanie Pivarski describe a module where students explore the environmental disaster caused by the catastrophic BP oil spill from an explosion on the Macondo well rig. The mathematical tools required include area computations using integration, and as such, the module is perfectly suited for a semester-long project in an integral calculus course, though Cohen and Pivarski offer suggestions for adapting it for a wide range of courses that cover differential equations and mathematical modeling. This is one of a handful of modules included in this volume exploring environmental issues; as argued eloquently by Jason Hamilton and Thomas J. Pfaff in their essay titled "Social Justice and Sustainability: Two Perspectives on the Same System", such issues are directly related to social justice. However Cohen and Pivarski also point out that working with real world data to investigate global corporations and their actions can empower students who might otherwise feel at sea in confronting faceless corporations which may seem invincible.

John Cullinan and Samuel Hsiao present a module suitable for discrete mathematics courses in their contribution titled "Voting with Partially-Ordered Preferences". With few mathematical prerequisites, the project may also be used for appropriate general education courses. Indeed the authors have used it in a course on mathematics and politics. The main mathematical ideas in the project are partial orders and voting theory. The central political question posed in the module is rather sophisticated and is bound to engage students. The design of the optimal political system is a complex question, and how best to represent voters' preferences is an unsolved (and perhaps unsolvable – see Arrow's Theorem mentioned in this module and elsewhere in the volume) problem. This module is one of a handful in this volume on elections and voting.

John Curran and Andrew Ross describe a role-playing game in their contribution titled "Implementing Social Security: A Historical Role-Playing Game". In this game, designed for a quantitative reasoning course, students reimagine and enact a fictional version of the historic debates in the United States Congress which eventually led to the passage of the 1935 Social Security Act. This project is an extensive one, and was originally developed for the Reacting to the Past (RTTP) consortium. In their contribution to this volume, Curran and Ross provide an overview of the project and the extensive resources freely available elsewhere for interested instructors. The module allows students to explore a complex social issue through the perspective of various constituents of a democracy and use mathematical tools to make convincing arguments to support their positions.

"Matching Kids to Schools: The School Choice Problem", by Julie Glass and Gizem Karaali, introduces a two-sided matching problem that confronts the challenge of finding fair ways to distribute students among the schools in a large school district, given constraints on student preferences and priorities such as proximity, special programs, and sibling placement. The module has no prerequisites and can be used in a course on discrete mathematics, or, with ample guidance, in a liberal arts mathematics setting. Throughout the project students are encouraged to consider what makes a matching algorithm fair and seek ways to encode these expectations in mathematical terms. Several exercises with solutions are provided.

In "Modeling the 2008 Subprime Mortgage Crisis in the United States", Bárbara González-Arévalo and Wanwan Huang describe a group project designed for a financial mathematics course. This module revolves around the 2008 mortgage crisis which caused an upheaval of the housing market in the U.S., affecting the lives and livelihood of many people. Students use real data to analyze the causes of the crisis and its effects on specific populations. The semester-long project ends with a poster session and involves students engaging with the local community, providing a neat way to incorporate civic engagement into a mathematics course.

Bárbara González-Arévalo and Wilfredo Urbina-Romero have designed a semester-long project for a second-semester calculus course in their contribution titled "Using

Calculus to Model Income Inequality". At the center is the Gini index, perhaps the most well-established measure of income inequality; in the module, students learn about the Gini index and then work to pose a good research question involving it. Studying the mathematics of the Lorentz curve, curve fitting, and numerical integration techniques, and collecting relevant data for their project, they develop the necessary expertise to investigate their research question. They wrap up the semester with a written report or a poster presentation. Directions that can be used to prepare suitable handouts are provided in the appendices. This is among a handful of modules revolving around the notion of income inequality.

Another contribution on elections and voting where Arrow's Theorem shows up is Kira Hamman's "What Does *Fair* Mean?" which describes a project that is specifically designed for a quantitative literacy course. In this module, students explore various voting methods and in the course of a class hour build up to Arrow's Impossibility Theorem. Hamman embeds the mathematical component of the project within a whole-class discussion of just what is fair. She provides suggestions for discussion prompts as well as an optional spreadsheet assignment for instructors who wish to explore the mathematics further.

Another module that draws attention to the social justice implications of environmental issues is "Social and Environmental Justice Impacts of Industrial Agriculture", by Amy Henderson and Emek Köse. In this extensive module, students work with mathematical models using ordinary differential equations to study the environmental impact of the waste created by industrial livestock farms across the United States. The material is suitable for courses in differential equations and mathematical modeling, and the authors provide handouts and instructor solutions for self-contained work, as well as resources which could support a multi-week project.

Reem Jaafar in "Student Loans: Fulfilling the American Dream or Surviving a Financial Nightmare?" presents a three-part module revolving around student debt, a topic that will be relevant to many college students. The main mathematical content focuses on polynomial models and their limitations. Therefore this project could be suitable for general education or quantitative literacy courses, as well as early parts of courses in mathematical modeling. The author provides a reading list for students and several handouts for the instructor on various components of the module.

Angela Vierling-Claassen's contribution titled "Modeling Social Change: The Rise in Acceptance of Same-Sex Relationships" introduces social network theory to explore social change. Students develop and then analyze a model that represents a social network where people's connections are strengthened or weakened progressively along with changes in their attitudes toward same-sex relationships. An optional computer-aided extension is also described. The mathematics involved is basic graph theory, and there are few prerequisites, so the module can provide a mathematical experience for students with widely differing backgrounds enrolled in discrete mathematics courses as well as general education or mathematics for liberal arts courses.

John Zobitz, Tracy Bibelnieks, and Mark Lester's "Sustainability Analysis of a Rural Nicaraguan Coffee Cooperative" grew out of an experience with a cooperative during a study abroad program and an Engaging Mathematics initiative, funded by NSF. The module exposes students to mathematical modeling in an introductory calculus course, motivated by researching the actual resources and concerns of a Central American coffee cooperative to promote sustainable practices in ecotourism. Students learn to apply ideas about rates, optimization, and modeling, with contextual resources including short introductory videos from the cooperative, guided homework analyzing basic revenue models, and the opportunity to analyze data from Google trends.

How to use this book

The essays in the book, which make up Part B, are organized in what seems to us a logical way. We expect that reading them in the order presented here will feel natural to most readers.

The modules, on the other hand, were harder to organize. We first thought about grouping them based on the courses they fit best. However as most modules would fit well into a range of courses rather than a specific single course, this ordering was somewhat imprecise. We considered the other thematic alternative, ordering the modules in terms of the socio-political issues they tackled. This too seemed somewhat ad hoc. In the end we grouped the modules into two separate volumes, each suitable for a range of courses, but with the more statistical or quantitative-reasoning approaches in *Mathematics for Social Justice: Focusing on Quantitative Reasoning and Statistics.*[4] We urge readers to skim through both books and be open-minded about what sort of mathematics will fit within what kinds of courses.

To help readers better navigate the contents of this volume, we have created two tables, presented in the following pages. The first table categorizes the modules by mathematical content and types of courses they would be appropriate for. The second aims to classify the modules in terms of the social justice issues that are explored within.

Our assignment of the modules into courses listed in Table 1.1 is merely suggestive. We hope that readers will explore all modules that might be mildly relevant to the class they are preparing to teach. Similarly, we have clustered the modules by social justice issue in Table 1.2, grouping the myriad issues into seven broad categories: *access, citizenship, environmental justice, equity/inequity, finance, human rights,* and *labor.* We hope that these tables will help readers get their bearings and feel comfortable exploring the contents of the volume.

While the tables provide an organizational guide to this eclectic collection of resources, we nonetheless believe that the best way to work with this book is by flipping through the pages, reading the abstracts, and skimming through the handouts in the appendices to see what might work for your classes. Through their efforts, the contributors to this volume have drawn compelling issues into the mathematics classroom and challenge us to do the same.

What next?

All our contributors are keen on sharing their work and will be happy to hear from you. If you are interested in any of the modules you find here and want to know more, or if you have any ideas about adopting one for your own context and would like to have access to source codes, TeX files, and so on, please feel free to contact the respective authors.

---

[4]See the two tables in the Postscript for more information on the content of *Mathematics for Social Justice: Focusing on Quantitative Reasoning and Statistics.*

We believe that this book is only a first step towards a broader goal: to establish an active and ongoing engagement of the college mathematics community with social justice issues. In order to instill the ideas of social justice into various introductory and general-education mathematics courses, we need to encourage the development of more teaching materials. It will take more effort for these kinds of concerns to influence the mathematics classrooms which center around more advanced material; we urge instructors to think creatively about how this can be done in courses such as abstract algebra, real analysis, and differential geometry. In the near future we hope to create an accompanying website which will contain up-to-date information about these modules, and links to resources and datasets. We hope to include in that site other modules as well, and we encourage readers to think about developing their own.

Last but not least, we need to bring to the broader mathematical community critical conversations about mathematics and its diverse and divergent roles in the social and public domains. Teaching mathematics for social justice is indeed much broader and more multifaceted than simply incorporating some social-justice themed modules into one's classroom. As we said: this is only the beginning.

A note on the cover art: The raised fist has a long history as a symbol of solidarity and of resistance against various forms of oppression. In this book cover, designed by Courtney Rose, the figure represents the possibilities of mathematics to empower us with a view towards creating a better world.

Table 1.1. Modules in this volume, categorized by relevant mathematical courses.

COLLEGE ALGEBRA

"Sea Level Change and Function Composition" by D. Archey (Chapter 7)

"Student Loans: Fulfilling the American Dream or Surviving a Financial Nightmare?" by R. Jaafar (Chapter 18)

GENERAL EDUCATION / LIBERAL ARTS

"Exploring the Problem of Human Trafficking" by J. Beier (Chapter 8)

"Evaluating Fairness in Electoral Districting" by G. Buhl and S. Q Kelly (Chapter 9)

"Voting with Partially-Ordered Preferences" by J. Cullinan and S. Hsiao (Chapter 11)

"Implementing Social Security: A Historical Role-Playing Game" by J. Curran and A. Ross (Chapter 12)

"Matching Kids to Schools: The School Choice Problem" by J. Glass and G. Karaali (Chapter 13)

"What Does *Fair* Mean?" by K. Hamman (Chapter 16)

"Modeling Social Change: The Rise in Acceptance of Same-Sex Relationships" by A. Vierling-Claassen (Chapter 19)

QUANTITATIVE REASONING

"Exploring the Problem of Human Trafficking" by J. Beier (Chapter 8)

"Evaluating Fairness in Electoral Districting" by G. Buhl and S. Q Kelly (Chapter 9)

"Student Loans: Fulfilling the American Dream or Surviving a Financial Nightmare?" by R. Jaafar (Chapter 18)

"Modeling Social Change: The Rise in Acceptance of Same-Sex Relationships" by A. Vierling-Claassen (Chapter 19)

CALCULUS

"Sea Level Change and Function Composition" by D. Archey (Chapter 7)

"Modeling the 2010 Gulf of Mexico Oil Spill" by S. Cohen and M. Pivarski (Chapter 10)

"Modeling the 2008 Subprime Mortgage Crisis in the United States" by B. González-Arévalo and W. Huang (Chapter 14)

"Using Calculus to Model Income Inequality" by B. González-Arévalo and W. Urbina-Romero (Chapter 15)

DIFFERENTIAL EQUATIONS

"Modeling the 2010 Gulf of Mexico Oil Spill" by S. Cohen and M. Pivarski (Chapter 10)

"Social and Environmental Justice Impacts of Industrial Agriculture" by A. Henderson and E. Köse (Chapter 17)

MATHEMATICAL MODELING

"Modeling the 2010 Gulf of Mexico Oil Spill" by S. Cohen and M. Pivarski (Chapter 10)

"Social and Environmental Justice Impacts of Industrial Agriculture" by A. Henderson and E. Kóse (Chapter 17)

"Student Loans: Fulfilling the American Dream or Surviving a Financial Nightmare?" by R. Jaafar (Chapter 18)

"Sustainability analysis of a rural Nicaraguan coffee cooperative" by J. Zobitz, T. Bibelnieks, and M. Lester (Chapter 20)

DISCRETE MATHEMATICS

"Exploring the Problem of Human Trafficking" by J. Beier (Chapter 8)

"Evaluating Fairness in Electoral Districting" by G. Buhl and S. Q Kelly (Chapter 9)

"Voting with Partially-Ordered Preferences" by J. Cullinan and S. Hsiao (Chapter 11)

"Matching Kids to Schools: The School Choice Problem" by J. Glass and G. Karaali (Chapter 13)

"Modeling Social Change: The Rise in Acceptance of Same-Sex Relationships" by A. Vierling-Claassen (Chapter 19)

GRAPH THEORY

"Exploring the Problem of Human Trafficking" by J. Beier (Chapter 8)

"Modeling Social Change: The Rise in Acceptance of Same-Sex Relationships" by A. Vierling-Claassen (Chapter 19)

GEOMETRY

"Evaluating Fairness in Electoral Districting" by G. Buhl and S. Q Kelly (Chapter 9)

COMBINATORICS

"Voting with Partially-Ordered Preferences" by J. Cullinan and S. Hsiao (Chapter 11)

CONTENT COURSES FOR TEACHERS

"Evaluating Fairness in Electoral Districting" by G. Buhl and S. Q Kelly (Chapter 9)

INTRODUCTORY STATISTICS

"Implementing Social Security: A Historical Role-Playing Game" by J. Curran and A. Ross (Chapter 12)

FINANCIAL MATHEMATICS

"Modeling the 2008 Subprime Mortgage Crisis in the United States" by B. González-Arévalo and W. Huang (Chapter 14)

Table 1.2. Modules in this volume, categorized by social justice theme clusters.

ACCESS

"Implementing Social Security: A Historical Role-Playing Game" by J. Curran and A. Ross (Chapter 12)

"Matching Kids to Schools: The School Choice Problem" by J. Glass and G. Karaali (Chapter 13)

"Using Calculus to Model Income Inequality" by B. González-Arévalo and W. Urbina-Romero (Chapter 15)

"Student Loans: Fulfilling the American Dream or Surviving a Financial Nightmare?" by R. Jaafar (Chapter 18)

CITIZENSHIP

"Evaluating Fairness in Electoral Districting" by G. Buhl and S. Q Kelly (Chapter 9)

"Voting with Partially-Ordered Preferences" by J. Cullinan and S. Hsiao (Chapter 11)

"Implementing Social Security: A Historical Role-Playing Game" by J. Curran and A. Ross (Chapter 12)

"What Does *Fair* Mean?" by K. Hamman (Chapter 16)

ENVIRONMENTAL JUSTICE

"Sea Level Change and Function Composition" by D. Archey (Chapter 7)

"Modeling the 2010 Gulf of Mexico Oil Spill" by S. Cohen and M. Pivarski (Chapter 10)

"Social and Environmental Justice Impacts of Industrial Agriculture" by A. Henderson and E. Köse (Chapter 17)

"Sustainability Analysis of a Rural Nicaraguan Coffee Cooperative" by J. Zobitz, T. Bibelnieks, and M. Lester (Chapter 20)

EQUITY / INEQUITY

"Using Calculus to Model Income Inequality" by B. González-Arévalo and W. Urbina-Romero (Chapter 15)

FINANCE

"Modeling the 2008 Subprime Mortgage Crisis in the United States" by B. González-Arévalo and W. Huang (Chapter 14)

"Using Calculus to Model Income Inequality" by B. González-Arévalo and W. Urbina-Romero (Chapter 15)

"Student Loans: Fulfilling the American Dream or Surviving a Financial Nightmare?" by R. Jaafar (Chapter 18)

HUMAN RIGHTS

"Exploring the Problem of Human Trafficking" by J. Beier (Chapter 8)

"Modeling Social Change: The Rise in Acceptance of Same-Sex Relationships" by A. Vierling-Claassen (Chapter 19)

LABOR

"Sustainability Analysis of a Rural Nicaraguan Coffee Cooperative" by J. Zobitz, T. Bibelnieks, and M. Lester (Chapter 20)

Bibliography

[1] Danielle Allen, *Our Declaration: A Reading of the Declaration of Independence in Defense of Equality*, Liverlight, New York, 2015.

[2] Mathew D. Felton-Koestler, Ksenija Simic-Muller, and José María Menéndez, eds., *Reflecting the World: A Guide to Incorporating Equity in Mathematics Teacher Education*

[3] Imani Goffney and Rochelle Gutiérrez, eds., *Rehumanizing Mathematics for Black, Indigenous, and Latinx Students*, National Council of Teachers of Mathematics, Reston, VA, 2018.

[4] Eric Gutstein, *Reading and Writing the World with Mathematics: Toward a Pedagogy for Social Justice*, Routledge, New York, 2006.

[5] Eric Gutstein and Bob Peterson, eds., *Rethinking Mathematics: Teaching Social Justice by the Numbers*, Rethinking Schools, Milwaukee, WI, 2005.

[6] Sara Hottinger, *Inventing the Mathematician: Gender, Race, and Our Cultural Understanding of Mathematics*, SUNY Press, 2017.

[7] International Council of Scientific Unions, *Millennium Ecosystem Assessment*, `http://www.millenniumassessment.org/en/index.html`, 2010.

[8] Robert Moses, *Radical Equations: Math Literacy and Civil Rights*, Beacon Press, Boston, 2001.

[9] Shannon Najmabadi, "Meet the Math Professor Who's Fighting Gerrymandering With Geometry", *Chronicle of Higher Education*, Volume 63, Issue 27 (March 10, 2017). Published online at `https://www.chronicle.com/article/Meet-the-Math-Professor/239260`, on February 22, 2017.

[10] `RadicalMath.org`, *Social Justice Math Resources*, online resource at `http://www.radicalmath.org/main.php?id=SocialJusticeMath`, accessed on April 7, 2017.

[11] Lynn Arthur Steen (ed.), *Mathematics and democracy*, National Council on Education and the Disciplines, Princeton, NJ, 2001. The case for quantitative literacy. MR1870000

[12] United Nations International Forum for Social Development, "Social Justice in an Open World: The Role of the United Nations", United Nations New York, 2006. Available online at `http://www.un.org/esa/socdev/documents/ifsd/SocialJustice.pdf`, accessed on May 5, 2017.

[13] Anita A. Wager and David W. Stinson, *Teaching Mathematics for Social Justice: Conversations with Educators*, National Council of Teachers of Mathematics, 2012.

Part B

Essays

2

Mathematics in Service to Democracy

Kira Hamman

In August 2005 I was a relatively new faculty member at a small liberal arts college in Maryland. Like virtually everyone else in the country, I watched in horror as Hurricane Katrina hit New Orleans. The devastation from the storm was appalling: nearly 2000 people dead, hundreds of thousands displaced, and over $100 billion in damage. But what was truly shocking was the scope of the poverty and inequality that was made evident in Katrina's aftermath. Before the storm, nearly 30% of the city's population, which was roughly two thirds African-American, lived in poverty. Almost a quarter did not have access to a car. As the levees crumbled and the water surged, we watched people try to escape the city on foot, only to be turned back by police. We watched them search desperately for food, water, and shelter as underprepared emergency facilities locked their doors. We watched the federal government, which seemed unaware of the scope of the disaster, fail spectacularly in its attempt to respond. And as we watched, our moral outrage grew.

My initial reaction was like everyone else's: this is horrible. But as I watched, that reaction changed into something a little different: why is everyone so surprised? We have decades' worth of data on poverty and racial inequity, most of it publicly funded and (at least in theory) publicly available. The poverty rate in New Orleans was readily available to anyone who cared to check, as was its relationship to race. The geographic position of the lowest-income neighborhoods and the consequent risk from flooding was hardly a secret. The lack of transportation and its effect on mobility in a disaster is well-documented. Why, then, was everyone so shocked?

And then it hit me: it was my fault. There I was, on the mathematics faculty at an institution of higher education. I was great at teaching students the chain rule and how to compute the volume of solids of rotation, but I had never taught anyone to use quantitative information to understand social issues. Nor, at that time, had anyone else I knew. Oh, maybe a statistician or two was using real data in the classroom, but they were the exception. Of course the country was surprised by what they saw in New Orleans in spite of the data. They didn't know about it because no one had ever taught them to think about it.

To be successful, a democracy requires educated citizens who are able to cooperate across differences to define, create, and maintain a just society. It requires citizens who are willing to participate in difficult decision-making and who take ownership of the knowledge necessary to inform those decisions. Perhaps most importantly, it requires that citizens believe in their own ability to affect change. Without those key ingredients – education, cooperation, participation – democracy falters. We see this with our democracy today, in which citizens perceive the political process as corrupt and their own participation in it as ineffectual. We have entered a kind of feedback loop: feelings of powerlessness lead to apathy and a lack of civic engagement and civic will, which in turn lead to disengagement from the knowledge and information needed to understand the big questions we face. Lack of understanding leads to further feelings of powerlessness, which lead to further alienation, and the cycle continues in a downward spiral.

Higher education has a crucial role to play in breaking this cycle. Indeed, many in higher education have already embraced this role, particularly in the social sciences. They aim not so much to figure out the solutions to the problems we face as to empower ordinary citizens to work together to find solutions. They acknowledge that real solutions are necessarily the result of participatory decision-making, and they seek to teach the social and organizational skills necessary for such participation. But that is not enough. These skills must have an underpinning of content knowledge so that the resulting decisions are based on evidence and reason.

This is where mathematics comes in. We live in an increasingly quantitative world, a world in which Google can as easily predict the pattern of a flu epidemic as choose which ads to display in your email. It's a world of unprecedented possibility, both positive and negative, and if we are to survive and thrive as a society we *must* have quantitatively literate citizens. The very real dangers of a quantitatively illiterate populace are regularly made apparent: outbreaks of whooping cough and measles due to imagined risks from vaccines; social policy based on anecdotal evidence; adjustable-rate mortgages. Furthermore, the explosion of technology has dramatically increased the number and significance of complex questions that are neither purely social nor purely quantitative. These questions exist between varied spheres of knowledge and understanding and cannot be addressed in the old ways. They require new ways of knowing and understanding, and new kinds of knowledge. Questions about things like genetic modification, artificial intelligence, and big data cannot, and should not, be answered by experts alone. We need the participation of ordinary citizens, but that participation is contingent upon those citizens' ability to contribute meaningfully to the discussion.

And so, after Katrina, I began to think about teaching mathematics differently. I discovered Lynn Steen's *Mathematics and Democracy* [4] and developed a course with that title, from which the module I contributed to this book is drawn. I hypothesized that students would be more invested in making democracy work if they could see what was under the hood, so to speak. I encouraged them to engage with, rather than shy away from, quantitative information in the media. I developed another course on that topic, modeled on John Allen Paulos's classic *A Mathematician Reads the Newspaper* [3]. If students could understand the data on things like poverty, health, and immigration, I thought, they would be much better placed to help society deal with those issues. In time, I came to believe that most students do not need to be creators of mathematics, instead they need to be savvy consumers of it. They need to be able to understand the world quantitatively.

Meanwhile, many other people in many other places were doing similar work. The National Numeracy Network had recently gotten its start, as had the Mathematical Association of America's Special Interest Group on Quantitative Literacy. While neither had a specific focus on social justice, both were driven by the idea that a healthy society needs citizens who are comfortable with quantitative information. Over the decade that followed, quantitative literacy became mainstream, and college mathematics curricula are the better for it. We no longer force students down the treacherous path to calculus when they have no interest in reaching its end. Most institutions offer classes like those described above, and publishers offer a range of resources to support them. We have come a long way.

But even as mathematics is part of the solution, I eventually came to understand that it is also part of the problem. Mathematics and social justice intersect in myriad ways beyond the college classroom, and not always positively. Outcomes in mathematics in the K-12 system are highly correlated to success in college and beyond, making the push for quantitative literacy curricula in the K-12 system a minefield in social justice terms. "Algebra for all!" cried Bob Moses in 1982. A former civil rights activist and high school mathematics teacher, he knew that middle school algebra acted as a gatekeeper for college admissions, yet the course was not even offered in many predominantly African-American middle schools. Moses wanted to make middle school algebra, and thus a shot at college, available to all students. And he was largely successful. Today algebra is, in most places, a high school graduation requirement.

The rub, of course, is that many both in and outside of the mathematics community have begun to argue that a pure algebra course for all students is at best silly, and at worst damaging to students without an interest in the STEM fields (see for example [1] and [2]). Despite the inflammatory rhetoric occasionally accompanying it, this is a defensible position, and one that is in line with the Steenian movement to bring mathematics education to bear on the problems of a participatory democracy. Those who expound this position do so with the intention of supporting civic engagement. But while their intentions are pure, the risks are great. The moment we track some students out of the algebra-calculus path, we recreate a situation which, in practice, has always disproportionately tracked low-income and minority students out of college and away from the STEM fields. This is the elephant in the room, and purity of intention does not grant us the right to look away from it.

Mathematics is a powerful tool for analyzing and addressing social injustice, but mathematics education is perhaps even more so. In 2005 in New Orleans, roughly 18% of African-American residents had a college degree, while 54% of White residents did. By 2015, the percentage for Whites had risen to 62%; the percentage for African Americans was still 18%. Before we can talk seriously about the power of mathematics to elevate democracy, we must address the issues of racial and social inequity that continue to plague our educational system. The moment we fall back on what *should be* rather than *what is*, we abdicate our responsibility as educators and as citizens. When we recognize and act on *what is*, both we and our students become the agents of change.

Bibliography

[1] Rick Gillman, *Reorganizing School Mathematics for Quantitative Literacy*, Numeracy: Vol. 3: Iss. 2, Article 7, 2010.

[2] Andrew Hacker, *The Math Myth and Other STEM Delusions*, The New Press, New York, 2016.

[3] John Allen Paulos, *A mathematician reads the newspaper*, Basic Books, New York, 2013. Paperback edition of the 1995 original with a new preface. MR3221808

[4] Lynn Arthur Steen (ed.), *Mathematics and democracy*, National Council on Education and the Disciplines, Princeton, NJ, 2001. The case for quantitative literacy. MR1870000

3

Preparing for Student Resistance: Rules of Engagement for Sensitive Topics

Lisa Marano

Recently I had the opportunity to teach several sections of our general education *Introduction to Applied Mathematics* course restricted to students in our Honors College. I was able to modify the content and delivery of the course from its typical lecture-driven format into one which incorporated significant amounts of in-class discussions.

The motto for the Honors College at West Chester University is: "To be Honorable is to Serve". To be true to this motto, I introduced supplemental materials on access to quantitative and financial literacy and fairness which developed students' empathy and awareness. Through readings, videos, and associated assignments, students' views and philosophical beliefs of historically marginalized groups were challenged. As a result, there was some student resistance, which manifested in several ways, from heated debates and teary eyed discussions in the classroom to unsympathetic or non-empathetic journal entries.

In this essay, I reflect upon encounters with and resolutions of student resistance, as this is a likely occurrence in a course of this type. I also discuss how a class can be organized to mitigate the effects of such encounters and reduce their occurrences.

1 Introduction

In April 2016, Donald Trump came to our University as part of his "Make America Great Again" tour. He quite literally divided our university in two along Church Street. On the east side of Church Street close to a thousand students, faculty, and staff rallied for hours chanting slogans like "Love Trumps Hate" and "Build Bridges Not Walls". A typical response from the west side of Church Street, "Build Them Higher", came from some of the thousands who stood in line to see Trump. The Trump supporters who waited for over four hours for the rally consisted mostly of locals. Some of our students did stand in line on the west side of Church Street as well.

During the afternoon, a group of students pushed a giant inflatable beach ball, "The Freedom of Speech Ball", around the crowds. People were encouraged to write on the ball whatever they were feeling at that moment. While on the east side, I saw some write, "Feel the Bern" and peace signs. I followed the ball to the west side of Church Street. Here, I witnessed a group of young, clean-cut white men wearing Trump t-shirts approach the ball. My heart sank as I witnessed one young man draw a swastika on the ball. I also witnessed a group of adult men ask a young black student if she was bused in for the protest. Encouraged by her friends, she countered by simply chanting again, "Love Trumps Hate".

This is happening at a time when comedians such as Chris Rock and Jerry Seinfeld refuse to come on college campuses. According to Seinfeld, he was warned not to "go near colleges—they're so PC" [2]. Rock stopped performing because of "their social views and their willingness not to offend anybody" [8].

There is a deep divide among some of our students and we must recognize this when dealing with social justice issues in our classes. It is essential that we plan for activities which will minimize student resistance while encouraging public discourse. In the sections which follow, I discuss how to generate positive classroom discussions to minimize student resistance.

2 The Course

On several occasions I taught a special section of our general education *Introduction to Applied Mathematics* course with enrollment reserved for students in our Honors College. Our Honors College strives to prepare students to become forces for positive change through several avenues including service and leadership. To reinforce the ideals and principles of the Honors College, I used material typically taught in the *Introduction to Applied Mathematics* course: mathematical finance, statistics, and probability, as the spine of the course. I integrated materials used in a *Mathematics and Social Justice Seminar* course where I focused on access to quantitative and financial literacy and to fairness. Many of these assignments were co-developed with Rob Root, Lafayette College, and Andy Miller, Belmont University, while we attended the Second Course Development Workshop in the Mathematics of Social Justice.

In this course, we explored several social issues and discussed methods which could quantitatively illustrate injustices. The hope was that each student would learn the mathematical skills and techniques required of an informed citizen. We often refer to this toolkit of basic mathematical skills as *Quantitative Literacy* (QL); see for instance [11].

More specifically the main objectives of this course were for students to:

(1) have a better understanding of Quantitative Literacy, and of how it can be used to analyze many social justice issues,

(2) be aware of the connection between quantitative literacy and the issues of fair treatment, and

(3) be aware of the increasing inequality of income and distribution of wealth in the U.S.

To build on these themes and reinforce concepts, we read excerpts from Lynn Arthur Steen's edited volume *Mathematics and Democracy* [12] and Elizabeth Warren's *The Two Income Trap* [14], and watched James Scurlock's documentary *Maxed Out* [10], for it is essential to ground the course on the close reading of high-quality texts [3]. In addition, the students were asked to engage in a service-learning project [7].

3 Strategies to Prepare Students for High-Quality In-Class Discussions

I assigned weekly readings from the texts and designed discussions to probe the students' views and philosophical beliefs. During the in-class discussion which followed, students were encouraged to share their thoughts.

The students had to be prepared for the discussions before they happened. We accomplished this by asking students to reflect on the readings in their journals via directed questions. Students were asked to relate the readings to their personal experiences; in addition, they would offer questions the readings generated. I found the journal entries which linked to their personal experiences were always the most authentic. Additionally, the journals were a part of their grade. They were graded on completion and thoughtfulness of responses.

For example, as one of the first assignments, the students were asked to read "The Case for Quantitative Literacy" [11]. Then they were asked to reflect on the following questions which served as a twist on a familiar mathematical biography assignment:

- *What do you think are your strengths as a quantitative thinker? Describe an experience where you used these strengths to make a decision.*

- *What do you see as your weaknesses? Describe an experience where your weaknesses were exposed or exploited.*

- *Describe an important learning experience in your own QL development.*

- *Have you ever had fun learning or using QL skills?*

Although this seems like a safe first assignment which allowed students to open up, it also gave them the opportunity to start talking about their similarities and differences in a low-stakes setting. It set the stage for future topics about which the students would find it more difficult to talk.

As this was a general education course, math-phobic responses were the norm. Students openly admitted that they were quick to use their phone to calculate something that could be easily done in their head, and many students realized that they were vulnerable to exploitation. They expressed fear about not knowing how credit cards worked and not knowing how to determine if their checking account balances were correct. However, I did get some surprising responses to the "fun" question. One of my favorite responses was from a student who worked as a server at a small chain restaurant. She writes,

> "Business was often slow, but I was a great waitress and often received fairly nice tips. However, being a server, I still made next to no money. So, I learned quickly how to calculate how many tips to claim. I would have to claim enough to show [the restaurant] that I was a good worker, but not enough that the restaurant would not compensate for a poor day (if our standard wage + tips did not equal $8/hour, they would compensate). It may sound dishonest, but, hey, I had to pay for school somehow!!! Sometimes I found this fun, because I reaped the benefits".

The take-away here is that in order to get the most out of in-class discussions, the students had to be prepared for discussions before class. To cultivate better discussions in the classroom, I had to design journal assignments which were linked to their high-quality readings and which asked the students to relate the reading to their own experiences.

4 Rules for Engagement

After students submitted journal entries to me using an online classroom management system, in-class discussion followed. Groundwork had to be set for this to happen successfully.

I employed a technique used during a semester-long workshop on diversity and inclusion in which I participated. The technique, designed by our diversity and inclusion specialists, was intended to ensure all participants felt comfortable speaking, especially those who may have felt traditionally silenced in classroom discussions; also it was intended to remind us that not everyone in class has the same level of comfort in speaking up and this comfort is a privilege typically enjoyed by the majority. The technique, a "no zing" policy, was used when a participant shared an opinion which differed from our own. We could challenge their opinion, but we could not react with a "zing" which was considered anything from a personal attack to a sarcastic response. The purpose of this policy is not to silence free speech, but to create a space where all feel safe to discuss their ideas. Conversations can and should get heated and opinions exchanged, but *personal* attacks would not be tolerated. I brought this "no zing" policy into my class. I reminded students that debate was about ideas, not personalities.

Once these ground rules were set, I managed the classroom so the discussion would not be dominated by the same voices throughout the semester. This was accomplished through my participation policy and organized opportunities for student engagement.

I let everyone know at the beginning of the semester that I would randomly call on everyone X times throughout the semester to share a part of their journal entry. X depended on class size and the number of discussion days we had. I used a random number generator to determine when each student would be called upon before the semester began. A typical class discussion began with me calling on three to four students and asking them to share part of their journal entry from the evening prior. If a student was absent or did not want to respond, they were allowed one free pass. Additionally, in class, every student had to respond to a classmate's shared journal at least twice each semester. As students were asked to share their journals and their class-mates responded, the socially engineered discussion began to take hold organically. Students were engaged and interactive. If after the organized discussion an idea from the readings did not materialize, I would ask for additional thoughts about those topics from the entire group. Finally, keeping the discussion on track was always important. On occasions, discussions tended to stray off topic. Responding in such a way that did not shut down students was important. "That is an interesting idea, but let's make sure we are focusing on the topic at hand", was a good way to steer conversation back [5].

Being flexible is still very important. In our class we watched the documentary *Maxed Out* where filmmaker James Scurlock examines the consumer-lending indus-try, in particular the practices of major banks, credit score reporting companies, and collection agencies. The deregulation of the banking industry is discussed, and some of the troubling practices of debt collectors and sub-prime lenders are highlighted. Signif-icantly, the film retells the stories of two college students who found themselves with insurmountable credit card debt. Their accounts are told by their mothers as both stu-dents felt the only way to escape their debt burden was by suicide. Also featured in the film is a grandmother who was harassed by debt collectors until she too took her own life.

The following discussion was the most powerful of my teaching career. That se-mester, I was teaching a class of twelve students. Of the twelve, three shared stories of friends or family they knew personally who took their lives because of credit card debt. In particular, one student's uncle had committed suicide earlier that semester be-cause of credit card debt due to medical expenses. Quickly, the planned discussion was thrown out the window, and we followed what the students needed. I reflected back a lot of what they were saying, using phrases like "I can hear the pain in your voice when you said…" and followed up with e-mail to certain students with information about the counseling center if they should be interested.

5 Assignments which Force Students to Appreciate Cultural Is-sues

In order to create a more effective learning environment for the students, I developed assignments which helped students "appreciate cultural issues at three levels:

(1) individual uniqueness;

(2) complex group identity, including intra-group differences, and

(3) those common human characteristics and behaviors we share across cultural and individual differences" [3].

For example, in one of the course assignments, students were asked about their idea of the "American Dream" as it pertains to personal finance and fairness. In par-ticular, they responded to the following questions:

What do the words "American Dream" mean to you?

How do you think your idea of the "American Dream" might differ from that of someone from a historically marginalized group?

In the next class, students worked in small groups to see how "alike" and "unalike" they were. Their directed discussion prompts asked them to:

(1) Share their "American Dream" ideas and uncover similarities and differences,

(2) Classify each American Dream: financial security, First-Amendment based, equal opportunity, etc.,

(3) Discuss how their response to 'What do the words American Dream mean to you?' may differ for students from a different race, class, sex, sexual orientation, age group, etc.

After the small group discussion, we reported back to the whole group what the most common similarities and differences were. Then I asked groups to share their responses to the third question and solicit input from other groups, asking if the other groups agreed or disagreed.

During the small group discussions, I overheard, on occasion, some biased ideas regarding immigrants and the LGBT community. However, it was easy in the small group setting for me to reflect back the statements made so that the students realized what they said was not appropriate and could correct their statements. Most often, the students did not realize they had said something insensitive or offensive and they appreciated not being "called out" in front of the entire class.

6 Preparing for Resistance

In courses which introduce issues of race, class, and gender, resistance from students can take three forms: vocal, silent, and absent [6]. Moreover, understanding the reasoning behind the resistance can be just as critical. Students may be resisting out of boredom. They might resist because the topic of discussion contradicts their own beliefs. Another reason may be low self-esteem regarding their academic abilities. However, "resistance and struggle ... can indicate a transformative learning experience" [13]. We must recognize that resistance is not necessarily a bad thing. I have used different strategies for each of the three forms of resistance.

Student journals[1] were my main tool to handle *silent resistance*. The journal assignments allowed students to share their views privately. I was able to provide students feedback and encourage them to think further on the issues so that over time, the students who chose to silently resist felt more comfortable speaking up in class. Using journals provided me with the opportunity to determine from where the resistance was coming. When students wrote short disengaged responses, students were typically either bored with the material or disagreed with the prompt. I would reflect on the discussion topic or respond to the journal entry by challenging the students' beliefs. Other times, it was obvious from the responses when students were afraid to out themselves as not being knowledgeable enough to discuss the topic adequately. From here, it is important to respond to the journal entries in a way which encourages students and fosters a culture that making mistakes is part of the learning process [13]. I provide an example of resistance as an opportunity of transformation next. I assigned the following journal prompt:

---

[1]Responding to journals has become increasingly easier through technology. Our course management system now allows us to respond to journal entries by leaving voice messages which reduces response times considerably.

In "Setting Greater Expectations for Quantitative Learning", Carol Geary Schneider describes her experience of the role of grade school mathematics as a tool for "sifting" and "sorting" students. Does this match with your own experience? If so, in what ways? Is this the most important role of mathematics education? What else might we hope for? Are there social justice implications to the use of mathematics as a tool for placing students on a scale? How does this choice affect girls/women? Minorities? Immigrants?

Responding to the latter part of this question, a student wrote:

"Also, immigrants may be placed lower or higher on the scale based upon where they are from. A new student from Mexico may not have taken as many rigorous math classes as a new student from Japan".

In class, this student did not contribute to the discussion even though it is clear that she has preconceived notions about Mexicans and Japanese. I wrote back to her in a way that was not accusatory, "I think that there are schools in Mexico which teach rigorous mathematics. What do you think? What do you think happens to immigrants, as a whole, who are taking English as a Second Language (ESL) in the sorting and sifting process?" This allowed her to rethink what she wrote and respond back in writing. Over time, this student began to open up more in class.

Some students will resist by not attending class on days when uncomfortable issues are being discussed; this is a good instance of *absent resistance* à la [**6**]. In my class, due to the random participation system that I used, students never knew when they would be asked to share their ideas. Missing class had the potential to hurt their grade if they were not present when they were randomly called upon to share their thoughts. As a result students rarely missed class.

Vocal resistance is the most difficult form of resistance to manage. A heated classroom discussion can quickly deteriorate when someone feels their assumptions are being challenged. As the instructor, it was essential for me to remind students that personalities are linked to ideas, and that feelings are connected to ideas [**3**]. I had a diverse group of students and encouraged discussion of differing points of view.

One time in class, to understand the students' beliefs regarding personal finance and fairness, I had assigned a two-part reaction paper. They were asked to react to an *Americans For Fairness in Lending* (AFFIL) publication, "Six Principles of Fairness in Lending" [**1**]. The six principles, *Responsibility*, *Justice*, *Equality*, *Information*, *Accountability*, and *Law & Enforcement*, were developed because the laws at the time allowed lenders to predatorily profit from personal debt. The students were asked to examine the six principles, and react to the following three questions:

- *If you had to eliminate one principle, which would you sacrifice? Why?*

- *Which, if any, of the principles alters your understanding of your personal relationship with credit? How so?*

- *How do these six principles protect historically marginalized groups?*

As with all reaction papers, there was class time allotted to discuss the student responses. A few students who self-identified as white middle-class men chose *Responsibility*, citing that individuals are responsible for their own financial situations. One student referred to an acquaintance who racked up credit card debt and filed for bankruptcy. The acquaintance continued to spend frivolously even after losing a job and while collecting unemployment. The student was angry that his family's tax dollars were continually paying to bail out people who could not get a job and were just irresponsible.

These students were employing a form of vocal resistance where they talked about the "exception". They were making generalizations about people who are in debt and collecting unemployment based on a few individual cases [6]. This solicited angry responses from other students. I had to remind the class of the rules for classroom discussion and that we criticize ideas and not people. Because of the random selection structure and the need to participate in class, I was able to include more students in the discussion than may have joined otherwise.

As long as the students stayed within the guidelines, I let the heated discussion continue until it stalled because I was prepared for this reaction. I had planned for the students to watch *Maxed Out* during our next class. After the discussion and the movie, the students were given the option to revise their reaction paper. After this exposé on the predatory nature of the personal finance industry, each student who was vocally resistant and who had chosen *Responsibility* changed their response. This was significant, as the revision was in fact a voluntary assignment. By using the highly effective supplement, *Maxed Out*, those who were the most vocal and ready to blame individuals for their situations were able to see another point of view clearly. They realized that for most people, debt and unemployment are most often not a result of laziness.

It can be hard for our students to see a world beyond their life experiences, especially if they come from a life of privilege and have not had any previous opportunities to examine the ideas of social structures and systems [6]. If as instructors we are willing to tackle such complex issues in our classrooms, we should be aware that it often takes some effort to help such students move beyond their original perspectives, but it is not an impossible task.

7 Conclusion

To reduce student resistance advance planning is required. Acknowledging that there will be differences in the classroom is the first step. From the start, rules for classroom discussion must be laid out so that all voices can be heard and respected. Ideas should be debated and criticized, not individuals. Planning high quality readings and supplementary materials is also essential. Without these, resisting students will find holes in any contrary arguments. Designing assignments which allow students to explore, contemplate, and rationalize their beliefs will create a more authentic experience for everyone in the class, including the instructor. Allowing opportunities for revisions after each classroom discussion is also important for measuring student growth. Assignments should also allow students to explore their sameness and differentness.

Recognizing and dealing with vocal, silent, and absent resistance are key. In mathematics, we may be luckier than in other disciplines when dealing with resistance. We can quantify injustices. We can show students that there is imbalance in the distribution of wealth in the United States by providing them with tools such as the Gini index.[2] We can support the ideas we are trying to convey with data. We must be prepared with the information when the challenge comes. Perhaps, if anything, the more difficult resistance we deal with is the resistance to mathematics, as our colleague from the

---

[2]EDITOR'S NOTE: See Chapter 15 for a module exploring the Gini index.

University of Pennsylvania who tried to do a little differential equations on an airplane and was flagged by a fellow passenger as a potential terror suspect can attest [**4**].

Bibliography

[1] Americans For Fairness in Lending (AFFIL) "Six Principles of Fairness in Lending", archived at `http://web.archive.org/web/20080816040127/http://www.affil.org/about/principles`, accessed on August 24, 2016.

[2] Flanagan, C. (2015) "That's Not Funny!" *The Atlantic*, September 2015. Available online at `http://www.theatlantic.com/magazine/archive/2015/09/thats-not-funny/399335/`, accessed on August 24, 2016.

[3] Frederick, P. (1995). "Walking on eggs: Mastering the dreaded diversity discussion". *College Teaching*, Vol. **43**/No. 3, pp. 83-92.

[4] Glusac, E. (2016). "Vigilance Gone Awry: When Math Gets Mistaken for Terrorism". *The New York Times*, May 18, 2016. Available online at `http://www.nytimes.com/2016/05/22/travel/american-airlines-passenger-terrorism.html`, accessed on August 24, 2016.

[5] Hadwin, Allyson and Susan Wilcox. (2001). *A Handbook for Teaching Assistants*. Kingston: Instructional Development Centre Publications, Queen's University. Available online at `http://ddi.cs.uni-potsdam.de/Lehre/WissArbeitenHinweise/HandbookTeachingAssistants.pdf`, accessed on August 24, 2016.

[6] Higginbotham, E. (1996). "Getting all students to listen: Analyzing and coping with student resistance". *American Behavioral Scientist*, **40**(2), 203–211.

[7] Marano, L., Dempsey, K., & Leiser, R. (2016). "Alternative Service-Learning Projects in Mathematics: Moving Away from Tutoring and Consulting". Forthcoming in C. Crosby and F. Brockmeier (Eds.), *Community Engagement Program Implementation and Teacher Preparation for 21st Century Education*, IGI Global, Hershey, PA.

[8] Rich, Frank. (2014). "Chris Rock Talks to Frank Rich About Ferguson, Cosby, and What Racial Progress Really Means". *Vulture*, November 30, 2014. Available online at `http://www.vulture.com/2014/11/chris-rock-frank-rich-in-conversation.html`, accessed on August 24, 2016.

[9] Schneider, C. G. (2001). "Setting greater expectations of quantitative learning". In L. A. Steen (Ed.), *Mathematics and democracy: the case for quantitative literacy* (Princeton, NJ: National Council on Education and the Disciplines), pages 99–106.

[10] Scurlock, J. (Producer/Director/Writer). (2006). *Maxed out: hard times, easy credit and the era of predatory lenders* [DVD]. United States of America: Trueworks.

[11] Steen, L. A. et al. (2001). "The Case for Quantitative Literacy". In L. A. Steen (Ed.), *Mathematics and democracy: the case for quantitative literacy* (Princeton, NJ: National Council on Education and the Disciplines), pages 1–22.

[12] Lynn Arthur Steen (ed.), *Mathematics and democracy*, National Council on Education and the Disciplines, Princeton, NJ, 2001. The case for quantitative literacy. MR1870000

[13] Vetter, Amy, Jeanie Reynolds, Heather Beane, Katie Roquemore, Amanda Rorrer, and Katie Shepherd-Allred. (2012). "Reframing Resistance in the English Classroom." *English Journal* **102**(2), pages 114–21.

[14] Warren, E. (2004). *The Two Income Trap*. New York: Basic Books.

4

Social Justice and Sustainability: Two Perspectives on the Same System

**Jason Hamilton and
Thomas J. Pfaff**

The situation appears untenable: Humans live on a big rock floating in space about 93,000,000 miles from the nearest energy source. The reason this works, of course, is that we exist in a highly complex system of interconnected parts that function to keep us alive. We rely on plants and other organisms to construct our atmosphere, regulate temperature within livable bounds, capture energy from the sun to feed ourselves, purify our water, and make soil. Humans rely on each other for goods and services, security, love, and psychological and emotional well-being. In short, we live in a complex social-ecological system: On this earth, there are no natural systems free of human influence, and there are no social systems free from the influence of the rest of the biosphere [19].

Despite the inextricable interconnections among the human and non-human parts of our system, those of us who want to make the world a better place have tended to gravitate toward one side or the other. Social justice activists are concerned about economic, political, and social inequities among people. Environmental activists are concerned about environmental quality and the harmful effects of human activities.

Given that our "life-support" system is, in fact, a social-ecological system, these two groups are ultimately working toward the same goal. Let us consider a simple analogy: A ship has hit an iceberg and is slowly sinking; on the ship a fire is burning and will soon hit the fuel tanks causing the whole ship to explode; a man on the ship is about to be murdered by a pirate. Which one of these issues is most important? Of course they all are equally important. Saving the man without putting out the fire is ultimately pointless. Stopping the ship from sinking does not help the man much if he is dead. Anybody who is trying to work to improve any part is working with everybody else to improve all the parts.

In many mathematics classes (as with most college courses in general), we tend to focus on a part of the system and seldom if ever discuss the social-ecological system as a whole. For example, it is easy to see how the tools of mathematics can be applied to climate change by considering temperature data sets, such as distributions of temperature changes across the planet, or working with simple climate models. It can be more difficult to consider the social effects of changing climate on human lives and well-being. Because of this difficulty, we mathematics instructors often shy away from the social part of the system and focus only on the model or the data. If we do choose to focus on a human issue, for example, by bringing into our classroom epidemiological data relating poverty to rates of childhood asthma, we often do not feel competent to use our mathematics to explore how this might be exacerbated or ameliorated by reinforcing or balancing feedback loops related to anthropogenic climate change.

Our goal in this essay is to demonstrate how mathematics can be used in the classroom to address linked issues in the social-ecological system that is the human life support system. First we show how the seemingly separate areas of social justice and environmentalism are brought together into a unified social-ecological system within the concepts of sustainability. Then we provide several examples of how to address the larger systems issues of sustainability in mathematics courses.

What is Sustainability?

How does the concept of sustainability help to overcome the false human/non-human dichotomy that humans have created? Despite the common conflation of sustainability with environmentalism, green purchasing, or energy management, it is a much richer concept. The meaning of the word *sustainability* is context-dependent. As shorthand for the term "sustainable development" [24], the concept of sustainability was developed as a new paradigm for improving overall human well-being by considering the coupled social-ecological system. Originally used only in the context of social change, sustainability explicitly referred to meeting the basic needs of all people and extending the opportunity to satisfy aspirations for a better life to everybody, with change being required in all countries, rich and poor alike.

What was the goal of sustainability? What was to be sustained? The sustainability of development itself. The point is that sustainable development is not a fixed endpoint or a steady-state condition of a group of variables with respect to time. Rather, it is a process of sustained change in which natural resource use, monetary investment, the orientation of technological development, and institutional change are consistent with future and present needs [23].

Given that the word sustainability is so often misused and abused it is no wonder that there is ambiguity in many people's minds regarding the relationship between it and other concepts such as environmentalism or social justice. Ultimately, the real goal is to adopt a whole system perspective where the boundaries between social justice and environmentalism fade and we gain the richness of a new, more integrated perspective.

Sustainability starts with four simple propositions, any of which can serve as an entry point for a mathematics course. We explore these below.

Thesis 1: We can do better. It is the best of times, it is the worst of times (apologies to Dickens). Humans are, on the whole, longer-lived, better nourished, more educated, and have a higher standard of living than at any other time in our history. At the same time, we certainly can, and must, do much better. Just look around: new and emerging diseases; climate instability; food and water insecurity; huge and growing wealth discrepancies; differential rates of poverty among different ethnic groups within countries; lack of access to quality education for many; and lack of gender equality [22]. Delineating these problems is not an indictment of our lifestyle or the decisions humans have made in the past. Rather, it is a realistic assessment that humans have come a long way and that we have a long way to go.

Thesis 2: We risk limiting the options of future generations. It is clear to scientists that the collective decisions and actions of humans today are negatively impacting the ability of future generations to meet their needs and aspirations. Global climate change [13], ozone destruction [25], degradation of ecosystem services [11], peak oil [2], accumulation of persistent toxins in the environment [20], new and emerging diseases [12], and trends in food production [5] all point to the same conclusion: human impacts on the planet are accumulating at a rate that endangers our well-being. Unless we agree to consign our children and grandchildren to a world of very limited potential, we must include issues of intergenerational equity into our planning and decision making. How do we keep environmental degradation from nullifying our advances in human rights? How do we keep poverty from being a driver of environmental degradation? Why are we in a position to even have to ask this question?

Thesis 3: The system is not simply "broken". Given this complex and multifaceted set of issues, how does one work towards improvement without becoming overwhelmed by the enormity of the task compared to the size of the contribution that one person (or even a group of people) can make? Traditionally, we have tended to subdivide such problems into categories that do not seem quite so daunting. For example, environmentalists tend to focus on environmental issues: air pollution, water pollution, habitat destruction, endangered species, etc. Social justice activists tend to focus on a different set of issues: wealth distribution, privilege, institutional racism, gender inequality, etc.

This division of problems into distinct areas is deeply imbedded in our thinking and the mental models we use to make sense of the world. It is also reflected in the way we govern our societies. For example, in the United States, the Department of Health and Human Services is a different entity from the Department of Labor, the Department of Agriculture, and the Department of Housing and Urban Development. The Environmental Protection Agency is a different entity from the Council of Economic Advisors, the Department of Commerce, and the Department of Energy.

These distinctions and divisions in our thinking are also embedded in our educational system. In K-12 education, social studies and government is taught as a distinct area separate from science. Mathematics is taught as a separate discipline from everything else. Even when we look at data or models of real world phenomenon, mathematics instructors typically will not even mention ethical, philosophical, political, or other areas of implication or connectedness. As an example, climate change is not only a scientific issue but also an ethical issue in that the people disproportionately impacted by climate change are not the ones causing the problem or benefiting from activities and processes causing the problem.[1] In higher education, Biology departments are separate from departments of Environmental Studies. Sociology is different from Psychology, which is different from Communication. The humanities are typically distinct from the social sciences, which are distinct from the natural sciences.

Perhaps a large part of the problem is the way we have chosen to address the problem. By breaking a complex system into distinct mental and physical entities, we have lost the ability to understand the relationships among the parts. Perhaps our systems are not "broken" but are functioning exactly as designed. Perhaps we have created the systems that inherently create our conflicts.

At issue here is that the social-ecological system that supports us physically, mentally, and emotionally is not simply complicated; rather, it is complex and cannot be meaningfully pulled apart and the parts studied in isolation [9]. Complicated systems are just simple systems with many parts. In simple systems, whether the parts are many or few, interactions among parts are well-defined and predictable, and thus the system as a whole is well-defined and predictable (at least in theory). This does not mean that the system or the problems arising from it are necessarily easy! For example, cars, photocopiers, and spacecraft are complicated systems. Complex systems consist of few or many parts, but the source and essence of complexity arises from the richness, intensity, and character of the interactions among constituent parts. Typically, these interactions lead to non-linear and/or emergent behavior (behavior that cannot be predicted by studying the parts of the system individually). Furthermore, the interactions (as well as the specific connections over which these interactions occur) constantly change, compounding the difficulty of thorough analysis by the formation/dissolution of amplifying/stabilizing feedback loops.

Thesis 4: There is no hierarchy of problems. We used the sinking ship analogy earlier to suggest that when dealing with systems, there is often no hierarchy of importance of problems. Either something is keeping the system from functioning as we wish, or it is not. In our world today, we have social problems, economic problems, and environmental problems. These are simply the problems that occur in our social-ecological system. One of the more pernicious debates we have engaged in as a society is, "Which is most important?" This discussion makes three fundamental assumptions, all of which are incorrect. First, that problems can be separated this way. Second, that there is a linear cause-and-effect relationship among these issues. And third, that a hierarchy can be established. To show the fallacy of this thinking, take the example of increased anthropogenic carbon dioxide in Earth's atmosphere. Is it an environmental problem? Yes, and it is a problem that has data that allow us to model it in mathematics classes. It is causing climate instability, habitat destruction, loss of species diversity, ocean acidification, desertification of susceptible habitats, sea level rise, etc., and some of these issues can be studied in a mathematics classroom. Is it an economic problem?

---

[1] EDITOR'S NOTE: Also see Victor Piercey's essay in this volume on quantitative ethics.

Yes. Higher food prices, strain on economic institutions, etc. Is it a social problem? Yes. We are seeing increasing numbers of climate refugees, loss of arable land for food production, and food and water insecurity. Furthermore, the most vulnerable of people, typically the poorest, are disproportionately affected by these changes. Yet typically most of this is not mentioned in a mathematics classroom when looking at increased atmospheric carbon dioxide.

Even with the intertwined nature of the issues associated with anthropogenic carbon dioxide, if there were direct linear cause-and-effect relationships among these issues, it might be possible to make headway by separating the issues into areas. But there are not. Does poverty drive environmental destruction or does environmental destruction drive poverty? Both. Does economic instability drive social instability or does social instability drive economic instability? Both. Does lack of access to quality education drive under-employment and un-employment or is it the reverse? Both. Because the causes and effects go both ways, there can be no strict hierarchy of importance or order to address the issues. Instead, they should be addressed simultaneously — and we should be making these connections when teaching.

Social Justice and Sustainability: The same starting point. Our discussion to this point thus lays out four fundamental assumptions or starting points for those of us who are considering the problems in the world:

(1) Humans can and must do better than this.

(2) The actions of people today are limiting the choices and options of our children and grandchildren.

(3) The system itself is the problem.

(4) In a complex world, problems cannot be approached on a reductionist one-by-one basis.

These four key statements, originally introduced in [8], might be a starting point for the earnest environmentalist. They might be the starting point for the dedicated social justice activist. They might be the starting point for most of the world's problems, but they are in fact the fundamental starting point for the young field of sustainability science [7]. The point we wish to emphasize is that social justice, environmentalism, and sustainability are all just different lenses on the same view: How do we work for improvements in human health, ecosystem health, societal health, and economic health? Literally, how do we make sustained improvements in the social-ecological system?

Examples

Now that we are thinking in terms of a system, let us look at a few issues from the two perspectives and see how the system might get discussed in a mathematics classroom. Our goal here is not to outline explicit lesson plans or problems, as is done in the many modules in this volume, but to see how we may provide a context for mathematical discussions of three sample topics.

Issue 1: Climate change. Climate change, at first glance, appears to be a purely environmental or sustainability issue. We can measure carbon dioxide levels in the atmosphere, calculate yearly average temperatures, and observe changing ice levels to name a few climate change related issues. From the perspective of a mathematics instructor, there are many ways to engage climate issues in the classroom. There are simple climate models as in Tung's book *Topics in Mathematical Modeling* [21], basic curve fitting [16], investigating melting sea ice [17], and distributions of temperature deviations [6], to name a few.

All of this can be done as if people did not exist at all. But when we are thinking of the social-ecological system the issues are much more complex than a "simple" study of a warming planet. It is people, predominantly wealthy, that use the majority of fossil fuels that emit carbon dioxide. However, a warming planet will disproportionately impact poor populations even though they are not the main contributors to carbon dioxide in the atmosphere.

Consider the peoples of the small Alaskan village of Newtok. The villagers of Newtok are often referred to as America's first climate refugees [27]. Rising temperatures and melting permafrost have led to the erosion of the banks of the Ninglick River to the point that the village has to be relocated before it is under water, and is, in fact, currently being relocated. These villagers are poor and did not contribute meaningfully to climate change, but they are being impacted. Who then pays the cost to relocate the village? Even moving the village is a complex socio-economic problem with the relocation having been halted at least once.

In this example, we set out a path where we got at the social justice issues by starting with the environment. It could certainly go the other way: Start by examining how an increase in the gas tax would impact people, which then might lead to how carbon dioxide emissions might change, and then a discussion on climate. Similarly, we could start the story by examining refugees around the world, which could lead to Newtok as well as other climate refugees. This brings us back to climate change. When we think in terms of systems, there is often no set starting point nor any clear and easy solutions.

In the end, to have projects in a mathematics class that only recognize the measurable environmental changes while ignoring the impact on peoples and the ethical issues involved is to ignore the social-ecological system. At the same time someone more focused on social justice concerns should not ignore the measurable changes to the planet. It is important to keep the focus on the entire social-ecological system.

Issue 2: Income inequality. Income or wealth inequality is a quintessential social justice issue.[2] There are numerous ways to incorporate this issue into a mathematics course. The topic lends itself nicely to introductory statistics with income and wealth distributions as excellent examples of skewed distributions. There are online resources such as the world top income database [1] and a now classic video on income distribution [18], calculus classes can discuss the Gini coefficient [4] as a measure of resource inequality including income and wealth. Many of these topics have time series associated with investigating changes of inequality over time.

At the same time, it is possible to consider income inequality as part of the larger issues within the framework of the social-ecological system. For example, the debate about hydraulic fracturing and drilling (fracking) in New York State ties this all together nicely. Those in favor of fracking were disproportionately lower on the economic ladder than those opposed to it. The reason can be easily summed up by the notion that

---

[2]EDITOR'S NOTE: See Chapter 15 in this volume for a module exploring this issue.

it is hard to worry about future generations when you cannot feed yourself today. As part of the framework we have noted that we are already limiting the options of future generations and a substantial piece of that problem is tied to income inequality.

Introducing some form of carbon tax to limit climate change is an equality issue also. Climate refugees now and in the future are disproportionately poor but yet are not the primary carbon emitters that caused the problem. Who should pay for their relocation? A more equal society could more easily spread out the cost of solving many of our ecological problems.

Issue 3: Lead exposure and crime. It is well documented that our judicial system impacts minorities much differently from non-minorities. Issues of race and the judicial system may appear to be a social justice issue but not part of our social-ecological framework. Given the abundance of quantitative data on the subject the issue is easily incorporated into statistics courses as well as other general education mathematics courses. How might this be part of the social-ecological system?

Overall crime rates have been declining for about two decades in the U.S. There are many explanations for this but they typically lack a universal explanation. Nevin has put forth the hypothesis that reduction in environmental lead, primarily from cars but also from paint, roughly two decades before crime started declining is at least part of the explanation [**14,15**]. Lead is an environmental pollutant with known impacts on human brain development and negative impacts on plant and wildlife populations [**26**]. Thus, lead is an ecological problem as well as a societal problem. It is a sustainability problem.

When young children are exposed to lead, the negative impacts on brain development increase their likelihood of making poor choices in their late teens and early twenties. Hence, there is a posited connection between lead exposure and crime. Nevin's work goes further and points out that inner city African-American populations were disproportionately impacted by environmental lead. The story on the lead, crime, and race connection is ongoing, but it provides an example of how an environmental toxin, lead, may have impacted communities two decades later while at the same time having a disproportionate impact on inner city residents. Given underfunded lead remediation efforts, income inequality is part of this story too, but recent gentrification in older neighborhoods has led to increases in lead exposure at all income levels. See [**4**] for more.

The Nevin data can be a resource for regression analysis, and is the topic of one of the modules in this book. This example is just one of many environmental justice issues.[3] Van Jones puts it succinctly in a 2008 interview when asked what stake people of color have in the environmental movement. He responded, "A big one. It's the people of color who are disproportionately affected by bad food, bad air, and bad water. People of color are also disproportionately unable to escape the negative consequences of global warming. Look at Hurricane Katrina. People of color need equal protection from the worst environmental disasters and equal access to the best environmental technologies" [**10**]. In many ways, Van Jones' ideas apply to not just people of color but to the economically disadvantaged.

---

[3]EDITOR'S NOTE: There are several modules in this volume exploring various examples of environmental justice. See Chapters 7, 10, and 17.

Conclusion

What is the difference between social justice and sustainability? Nothing. Nothing, that is, if we remember that humans live in a coupled complex social-ecological system. It is the system itself that allows us to exist on this earth. It is changes in the system itself that social justice activists and sustainability advocates are promoting. And we are all working toward a common goal: improved physical, psychological, and emotional well-being for all.

Bibliography

[1] Facundo Alvaredo, Anthony B. Atkinson, Thomas Piketty, Emmanuel Saez, and Gabriel Zucman. *WID-The World Wealth and Income Database*, http://www.wid.world/, accessed on September 18, 2016.

[2] Association for the Study of Peak Oil & Gas, website at http://peak-oil.org, accessed on September 18, 2016.

[3] Michael T. Catalano, Tanya L. Leise, and Thomas J. Pfaff. "Measuring Resource Inequality: The Gini Coefficient." *Numeracy* 2(2): Article 4. 2009, 1–24. Available at http://services.bepress.com/numeracy/vol2/iss2/art4/, accessed on September 18, 2016.

[4] Kevin Drum. "Lead: America's real criminal element." *Mother Jones*, web content posted February 11, 2016. Available at http://www.motherjones.com/environment/2016/02/lead-exposure-gasoline-crime-increase-children-health, accessed on September 18, 2016.

[5] Earth Policy Institute. *Eco-Economy Indicators: Grain Harvest*. Web content available at http://www.earth-policy.org/indicators/C54, accessed on September 18, 2016.

[6] NASA Goddard Institute for Space Studies, *GISS Surface Temperature Analysis* (GISTEMP). Web content available at http://data.giss.nasa.gov/gistemp/maps/, accessed on September 18, 2016.

[7] J. G. Hamilton. "Plant Ecology and Sustainability Science", in *Ecology and the Environment: The Plant Sciences* **8** edited by R. Monson, (Springer, 2014), pages 631–654.

[8] J. G. Hamilton and T. J. Pfaff. "Sustainability Education: The What and How for Mathematics". *PRIMUS* **24**(1) (2014), 61–80.

[9] J. G. Hamilton, T. J. Pfaff, M. Rogers, and A. Erkan. "On Jargon: 21st-Century Problems. *The Journal of Undergraduate Mathematics and its Applications* **34**(4) (2013) 327–335.

[10] D. Kupper. "Bridging the Green Divide: Van Jones on Jobs, Jails, and Environmental Justice". *The Sun Magazine*, Issue 387, March 2008. Web content available at http://thesunmagazine.org/issues/387/bridging_the_green_divide?page=1, accessed on September 18, 2016.

[11] Millennium Ecosystem Assessment Board, *Living Beyond our Means: Natural Assets and Human Well-Being*. Report dated March, 2005, available at http://www.millenniumassessment.org/documents/document.429.aspx.pdf, accessed on September 18, 2016.

[12] National Institute of Allergy and Infectious Diseases. Emerging and Re-emerging Infectious Diseases. Retrieved originally from http://www.niaid.nih.gov/topics/emerging/Pages/Default.aspx, site content, dated April 22, 2016, archived at Internet Archive at https://web-beta.archive.org/web/20160422215043/http://www.niaid.nih.gov/topics/emerging/Pages/Default.aspx.

[13] National Snow and Ice Data Center. *Sea Ice Index*. Data set available at http://nsidc.org/data/g02135.html, accessed on September 18, 2016.

[14] R. Nevin. "How Lead Exposure Relates to Temporal Changes in IQ, Violent Crime, and Unwed Pregnancy". *Environmental Research*, **88** (2000) 1–22.

[15] R. Nevin. "Understanding International Crime Trends: The Legacy of Preschool Lead Exposure". *Environmental Research*, **107** (2007) 315–336.

[16] T. Pfaff. "Educating Students about Sustainability while Enhancing Calculus". *PRIMUS* **21**(4) (2011) 338–350.

[17] T. Pfaff, A. Erkan, M. Rogers, and J. Hamilton. "Multidisciplinary Engagement of Calculus Students in Climate Issues". *Science Education and Civic Engagement, An International Journal*, Winter 2011, 52–56. Available at http://www.seceij.net/seceij/winter11/calculus_studen.html accessed on September 18, 2016.

[18] Politizane. *Wealth Inequality in America*. Video available at https://www.youtube.com/watch?v=QPKKQnijnsM, accessed on September 18, 2016.

[19] C. L. Redman, J. Grove, and L. H. Kuby. "Integrating Social Science into the Long-Term Ecological Research (LTER) Network: Social Dimensions of Ecological Change and Ecological Dimensions of Social Change". *Ecosystems*, **7**(2) (2004), 161–171. Available at http://www.jstor.org/stable/3658606, accessed on September 18, 2016.

[20] S. Steingraber, *Living Downstream: An Ecologist's Personal Investigation of Cancer and the Environment*. Da Capo Press, Cambridge MA, 2010.

[21] K. K. Tung, *Topics in mathematical modeling*, Princeton University Press, Princeton, NJ, 2007. MR2311847

[22] United Nations Educational, Scientific, and Cultural Organization (UNESCO). *Education for Sustainable Development*. Web content archived at http://www.unesco.org/new/en/education/themes/leading-the-international-agenda/education-for-sustainable-development/, accessed on September 18, 2016. Also see the updated source at http://en.unesco.org/themes/education-sustainable-development, accessed on September 18, 2016.

[23] United Nations Educational, Scientific, and Cultural Organization (UNESCO). *UNESCO and sustainable development*. Report dated 2005. Available at http://unesdoc.unesco.org/images/0013/001393/139369e.pdf, accessed on September 18, 2016.

[24] United Nations General Assembly. *Report of the World Commission on Environment and Development*. 96th plenary meeting, December 11, 1987. Available at http://www.un.org/documents/ga/res/42/ares42-187.htm, accessed on September 18, 2016.

[25] United States Environmental Protection Agency. *Ozone Layer Protection*. Web content available at http://www.epa.gov/ozone/science/, accessed on September 18, 2016.

[26] United States Geological Survey (USGS). *Concerns Rise Over Known and Potential Impacts of Lead on Wildlife*. Web content available at http://www.nwhc.usgs.gov/disease_information/lead_poisoning/, accessed on September 18, 2016.

[27] W. Yardly. "Victim of Climate Change, a Town Seeks a Lifeline". *The New York Times*, May 27, 2007. Available at http://www.nytimes.com/2007/05/27/us/27newtok.html?pagewanted=all, accessed on September 18, 2016.

5

Quantitative Ethics

Victor Piercey

In late March, 2014 there was pressure on the Obama administration. The administration had predicted that enrollment with `https://www.healthcare.gov` would reach 7 million by the end of the month. The opposition press was gleefully sharing a graph which suggested that enrollment was well below expectations. However, there was a problem with the graph. The vertical axis met the horizontal axis at a point considerably above the origin. The resulting bar graph distorted reality. In response to critics, the bar graph was revised and rereleased. The revised graph told a dramatically different story — that although still short, the enrollment numbers were quite close to the projected goal.

In most quantitative literacy or quantitative reasoning courses, the mathematics courses where this issue might come up, the take-away lesson from this experience is to read graphs carefully and skeptically. This is a valuable lesson. However, it is limited in that it treats students as consumers of mathematical information. Implicitly, we are telling our students that they will confront quantitative information that is misleading and it is our job as informed citizens to see through these arguments.

The fundamental premise of what I call quantitative ethics is to turn this idea on its head. Rather than treat students as consumers of mathematical information, some students should be treated as producers of mathematical information. For example, business students and journalism students are ultimately going to create quantitative stories for their colleagues and readers. Rather than asking such students to appreciate that a graph can be misleading, I ask them whether it is appropriate to do so and why. This is an ethical question concerning quantitative information — quantitative ethics.

Roughly speaking, quantitative ethics asks students to consider the moral and societal implications of how we use data and other quantitative information. These are questions students typically have not encountered. The idea occurred when I asked students in a class whether they thought it was reasonable for somebody to select data that supports a predefined conclusion and ignore any other data. Most students thought that this was OK, indicating to me that students did not connect choices concerning data with real obligations to the public.

The idea of quantitative ethics was motivated by my former career as an attorney. As an attorney, my job was to use my expertise to advise clients and make recommendations. The legal profession, like all other professions, is bound by a code of ethics. A code of ethics is appropriate whenever knowledge and expertise puts one person in a position of authority over another. As I listened to my students justify cherry-picking data to fit a desired conclusion, it occurred to me that an ability to read and interpret data, particularly specialized data, or having access to non-public data, puts one in a similar position of authority as any other professional. Why shouldn't the public expect that authority to be handled responsibly?

I am developing the idea of quantitative ethics inductively. This means that I am considering problems and questions and developing a framework and definition in response to the examples. At present, I have identified five components of quantitative ethics, three of which I have used in the classroom. These five components are:

Decisions: How can quantitative information be used to make more ethical decisions? What moral and ethical dilemmas arise when quantitative information is part of our decision-making process?

Communication: What ethical or moral guidelines should we follow when communicating quantitative information and summarizing data?

Assumptions: What assumptions go into our quantitative models? What ethical considerations should be made when selecting those assumptions and what are the consequences?

Framing questions: How do ethical and moral implications arise from the way we frame our questions?

Serving clients: How do we handle a situation when the data we are faced with yield conclusions undesirable to a client?

The first three are based on questions that I have incorporated into the classroom. The other two can also lead to rich classroom questions and discussions, but I have not done so yet.

To date, I have incorporated quantitative ethics into my quantitative reasoning course sequence for business students. This is a contextualized, inquiry-based sequence of two courses. The course was designed for university students with business majors who place below the college level in mathematics. I drafted the course materials[1] as "Explorations" — guided problems that yield mathematical meaning. Unfortunately, time constraints have limited my ability to use the quantitative ethics questions that I have drafted, but we have been able to discuss some. In upcoming revisions I will make room to dedicate to ethical discussions.

---

[1]I would be happy to share these materials with readers who reach out to me via email.

There are two ways quantitative ethics is weaved into the course materials. The first way involves "Focus on Quantitative Ethics" — specific ethical questions embedded in explorations and related to the mathematical content or business context in that exploration. The other approach involves "Case Studies in Quantitative Ethics" — entire explorations dedicated to a recent business instance that raises ethical questions.

A "focus" question related to decisions has to do with credit card minimum payments. Several explorations are designed to guide students to solving "the Credit Card Problem". The Credit Card Problem asks, given a fixed initial balance P on a credit card, a fixed interest rate r compounded monthly, and a fixed monthly payment M, how long does it take to pay off a credit card? The solution is given by the equation:

$$n = \frac{\ln\left(\frac{M}{M-rP}\right)}{\ln(1+r)}.$$

Such a problem fits very naturally into a unit on logarithms. In the process of working through the problem, a need for logarithms naturally arises!

The solution to the credit card problem raises an interesting mathematical question related to the domain of the logarithm. The denominator has no problems since $1+r > 0$ as $r > 0$. However, the numerator may not always be defined. Namely, in order for the logarithm to be defined, we need to have $M > rP$. This makes intuitive sense, it tells us that in order to be able to pay the credit card balance off in finite time, we need our monthly payment to be larger than the dollar amount of interest accumulated on the principal each month.

The ethical question here has to do with a decision regarding minimum payments. Specifically, is it ethically appropriate to offer a credit card holder the option to pay a minimum monthly payment that is less than rP? Many consumers who find themselves in credit card debt and are struggling to make ends meet will only make the minimum monthly payment in order to stay current and stay afloat. Mathematically, when $M \leq rP$, the solution to the credit card problem flies to infinity. With equality, the balance stays constant, and with strict inequality, the balance continues to rise. As a decision-maker with a bank, in the absence of government regulations, do we allow our customer to fall deeper into debt while believing they are paying it off? Assuming the customer makes their payments on time, this guarantees the bank a regular stream of income. But this comes at an ethical cost in that we are manipulating our borrowers.

Of note is how this question is different from how this problem would be associated in other courses. Typically, we assume our students will be credit card users — consumers in the scenario. The lesson to the consumers is that one should be wary of minimum credit card payments and actually how to calculate a reasonably monthly payment plan. This is a valuable thing to learn — for all students! However, the ethics question puts the student in a decision-maker's place. Now they have to use quantitative information to make a decision that affects real lives every day! It is also worth noting how this scenario arises out of the domain of the logarithm function, an otherwise very dry and abstract concept that I would be tempted to ignore in a standard general education mathematics course.

Another focus question, this time having to do with communication, concerns how interest rates on credit cards or loans are quoted to consumers. The truth-in-lending law requires that the nominal interest rate, the APR, be disclosed to borrowers. However, unless interest is compounded annually, this is not the actual interest rate accumulated annually. When we account for compounding over the course of a year, the effective annual interest rate — EAPR — is actually higher than the quoted APR. If the interest compounds m times per year (what I call the compounding frequency), then we can calculate the EAPR using the formula:

$$EAPR = \left(1 + \frac{APR}{m}\right)^m - 1.$$

The difference can be quite significant. If the quoted APR is 24%, the effective APR is 26.8%.

In a unit on solving and manipulating formulas, I ask students to solve this formula for APR. They are then given an Excel spreadsheets with many effective APRs and compounding frequencies and asked to program their solved formula into Excel to calculate the nominal APR in each case. This is a straightforward exercise. The mathematical problem certainly challenges students while the Excel component helps sell the students on the utility of solving a formula for a specified variable. The ethical question arises when we ask why we would want to do this.

The scenario we imagine is a bank manager who has targeted values of the effective annual rate that they would like to collect along with the compounding frequency, and trying to calculate the nominal APR they will quote to consumers in the required truth-in-lending disclosures. Is this appropriate? Is this limited communication misleading? One can imagine a regulator or a legislator considering this question. An interesting assignment for students would be to ask them to write a letter to their Congressional representative asking them to consider this very issue, since neither the compounding frequency nor the effective annual rate are required to be disclosed under the truth-in-lending law.

There are also focus questions I have written concerning the assumptions we make in our mathematical models. At one point in the course, students are asked to solve a linear programming problem involving how many stores to open in a particular area.[2] The company has three types of stores: convenience stores, standard stores, and expanded service stores. There are costs to support each type of store, projected revenue streams from each type of store, a fixed number of employees required for each type of store, along with an overall budget constraint and a limit on the total number of employees. The task is to select the numbers for each type of store that will maximize profit.

The assumptions in this problem have to do with cost and predicted revenues. Note that the costs are all pecuniary — they involve construction costs and operating costs. However, they ignore community costs. What will the impact on traffic be? What about the amount of green-space per person? What about environmental impacts? What about the population? The ethics question is whether these should be considered. If one were on a local government board tasked with approving such projects, how does one balance these costs with the potential for jobs in the community? If one were a manager for the company, how does one balance the obligation to maximize profit for the shareholder with moral and ethical obligations to the community? From the perspective of mathematical modeling, one can attempt to include these costs into

---

[2]EDITOR'S NOTE: Forest Fisher and Jared Warner in their module "A Gentrification Module for Quantitative Reasoning" in *Mathematics for Social Justice: Focusing on Quantitative Reasoning and Statistics* pose this question, along with a few others.

the model. From an ethical perspective, are these the costs that we should consider in our assumptions?

The current version of my course sequence includes three case studies on quantitative ethics. The first case study, which appears in a "reflection week" to end the first course in the sequence, concerns the Enron Corporation, the energy, commodities, and services company whose 2001 demise is today upheld as a striking example of corporate fraud. The best reference for this story remains *The Smartest Guys in the Room: The Amazing Rise and Scandalous Fall of Enron* by Bethany McLean and Peter Elkind [1].

The Enron failure concerned two choices that have to do with how data are communicated to stakeholders. The first choice was to hide debt in special-purpose entities. These were subsidiaries created for the purpose of assuming Enron debt associated with various deals, and were designed in a way that allowed Enron to avoid reporting their existence and their financials on their accounting statements, but preserved the ability of Enron executives to effectively control them.

The second choice was to use mark-to-market accounting to report income. Mark-to-market accounting is a reporting mechanism that is typically used for investments and differs from cost accounting. To illustrate, suppose a company purchases a share of stock for $20 and during the quarter the stock's value rises to $30. Under cost accounting, if we have not sold the stock, we do not report any income and value the stock on our balance sheet at $20, what we paid for it. Under mark-to-market accounting, we use the $30 value of the stock on our balance sheet and report $10 in income even if we have not sold the stock yet. Of course, if the stock loses value we would report a loss.

Mark-to-market accounting arguably makes sense for investments whose values fluctuate. However, Enron used them to report income from deals that they made. For example, at one point in the late 1990s, Enron cut a deal with Blockbuster to stream video games. In their accounting statement, using mark-to-market accounting, Enron reported as income the present value of the future expected income stream from this deal. This was how Enron applied mark-to-market accounting: they would make a deal and immediately report the income from the deal on their balance sheet.

There are several quantitative ethics issues arising out of this scenario. Did Enron appropriately communicate quantitative information to their stakeholders? Less obvious, even if these communication habits were ethically appropriate, were these decisions ethically sound? Consider the incentive structure these decisions create. If one reports all of the income from a deal during the period in which the deal closes, that income cannot be reported in later periods. This means that in order for the books to reflect a financially healthy company, executives need to continue to cut more and more large scale deals. Since the debt is buried off-book, the deals may not be financially sound. Moreover, the application of mark-to-market accounting to these deals assumes that the income will materialize. This assumption itself can be ethically questioned. The Blockbuster deal, for example, never worked out as the technology was not available at the time.

Another case study concerns Ponzi schemes. This takes place in a "preview" week for the second course, in which we explore themes associated with mathematical modeling and prediction. In a Ponzi scheme, an operator who originates the scheme offers an investment opportunity with an unusually large return on investment. Once investors who are interested buy in, the operator goes on to another level wherein a larger investment with extravagant returns is offered. The original investors are paid off using the proceeds from the second level, and the scheme continues to iterate. In the case study, students are asked to construct a numerical and visual model of a Ponzi scheme. They use this model to predict why Ponzi schemes will ultimately fail. Ponzi schemes are contrasted with multi-level-marketing arrangements in which product sales are involved.

Modeling Ponzi schemes and comparing to multi-level marketing raises several ethical questions. As legislator, how should government intervene? Should either of these arrangements be outlawed? Why? Is it more appropriate to expect that the buyer should beware? Should investors self-govern? How does the mathematical model predicting failure at least for Ponzi schemes impact your decision? How should the operator ethically communicate the investment opportunity? If one can construct a mathematical model in which a Ponzi scheme does not fail, what assumptions are involved and are they realistic?

A third and final case study concerns the 2008 economic collapse. Based on the opening chapter of Nate Silver's *The Signal and the Noise* [2], we look at the role ratings agencies played in the run-up to the recession. The exploration occurs in the final "reflection week" in the second course. Since the second course is about mathematical modeling and prediction, this fits Silver's accusation that the financial collapse arose from "a failure of prediction".

The recession was a result of the failure of the mortgage-backed securities market. The idea of mortgage-backed securities is that one takes a large pool of mortgages (or any other set of loans or income-producing assets) and offers to the public securities that pay shares of the income from the mortgages. The process is called "securitization". As with other securities issued to the public, ratings agencies evaluate such investment products.

When assessing the risk of a mortgage-backed security, one must assess the probability that the mortgages result in foreclosure. This requires a stochastic model, one which requires assumptions. One critical assumption made by the ratings agencies was that the probability of foreclosure of one mortgage was independent of the probability of foreclosure of any other mortgage in the pool. This assumption is unrealistic. When one house in a neighborhood goes into foreclosure, it will impact the value of neighboring homes. Moreover, a foreclosure may indicate a local economic problem such as a series of layoffs at a large regional employers. Silver illustrates pointedly that the independence assumption results in a materially different evaluation.

Was this assumption ethically warranted? If not, this case study raises the question when an assumption in a mathematical model is ethical. Some assumptions are always required, otherwise a modeling problem may not be tractable. Moreover, this case raises the question of how the independence assumption — and its consequences — should be communicated to the investing public. Is this standard different when the investor is a more sophisticated operator such as Lehman Brothers?

The examples above illustrate that a mathematical classroom is fertile ground for interesting ethical questions, questions concerning "quantitative ethics". In my future classwork, I plan to prioritize the case studies and focus questions by making room. In addition, I think it is important to develop problems that do not have a resolution that the students will find obvious. We need to explore ethical problems where there are strong arguments on each side. Finally, these problems need to be designed to be more accessible to my student population. For example, most first-year college students to-day have not heard of the Enron scandal, which was a very complicated and involved technical issue. One way I am attempting to accomplish these changes in the next re-vision of the course is embracing "game-based learning". For example, I am planning on developing a week-long Enron simulation which will involve several course objec-tives.[3]

Notwithstanding the necessary improvements, quantitative ethics is an idea with merit that has great potential. I include in the Appendix some student work from a fo-cus question. In this sample, we see the student thinking carefully about the real-world consequences of a decision concerning stock valuation. Quantitative ethics discussions help connect math to the real world and illustrate to students that math is not always "black and white". There is also great potential to develop interdisciplinary collabo-ration with other faculty who have an interest in ethics. Finally, there is always the hope that by infusing ethics across the curriculum, we might be able to improve future ethical behavior.

---

[3]EDITOR'S NOTE: There are several modules in this volume that use role-playing as a significant peda-gogical tool. See Chapters 12 and 19.

Appendix: Student Work Sample

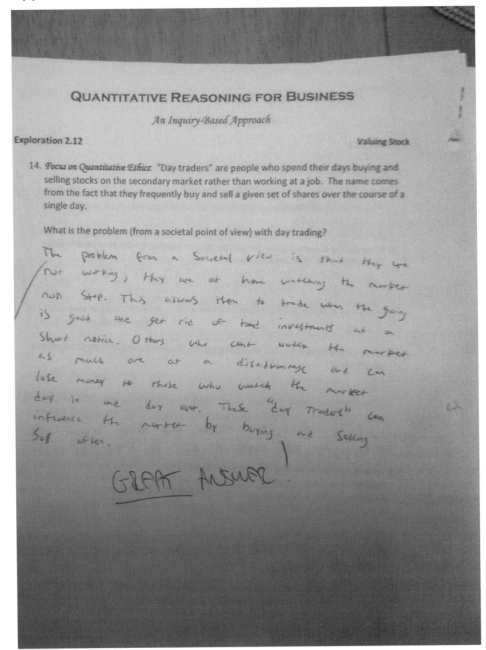

QUANTITATIVE REASONING FOR BUSINESS

An Inquiry-Based Approach

Exploration 2.12   Valuing Stock

14. *Focus on Quantitative Ethics.* "Day traders" are people who spend their days buying and selling stocks on the secondary market rather than working at a job. The name comes from the fact that they frequently buy and sell a given set of shares over the course of a single day.

What is the problem (from a societal point of view) with day trading?

The problem from a societal view is that they are not working, they are at home watching the market run stop. This allows them to trade when the going is good one get rid of bad investments at a short notice. Others who cant watch the market as much are at a disadvantage and can lose money to those who watch the market day in and day out. These "day Traders" can influence the market by buying and selling Sof often.

GREAT ANSWER!

Bibliography

[1] Bethany McLean and Peter Elkind, *The Smartest Guys in the Room: The Amazing Rise and Scandalous Fall of Enron*, Portfolio Trade, New York, 2003.

[2] Nate Silver, *The Signal and the Noise: Why So Many Predictions Fail–But Some Don't*, Penguin Press, New York, 2012.

6

Math for Social Justice: A Last Math Class for Responsible Citizens

Dave Kung

The fall of 2008 marked my sixteenth year teaching college students mathematics. When I had taught Survey of Mathematics, our liberal arts math course, I had always struggled to win non-STEM hearts by gently guiding the reluctant students toward the most easily accessible vistas overlooking our beautiful subject. The impetus to change how I approached this "dead-end" course, likely the last math class these students would ever take, came from an unlikely source: the Obama-McCain presidential debates.

Senator McCain: "*Senator Obama's secret that you don't know is that his tax increases will increase taxes on 50% of small business revenue*".

Senator Obama: "*Only a few percent of small businesses make more than $250,000 a year. So the vast majority of small businesses would get a tax cut under my plan*". "*98% percent of small businesses make less than $250,000*".

As a quick-thinking math geek, it didn't take me more than a second to reconcile these seemingly contradictory statistics. Yes, one is talking about revenue, the other profits — and they might have different definitions of the size of a "small" business — but the core of the issue was a simple fact about distributions: a huge portion of small business revenue (50%) might go to a small portion of the businesses (2%).

Seconds later, my self-satisfied smugness about quickly solving this political puzzle faded to depression. This little nugget of mathematical reasoning, vital for understanding the national dialogue, was a skill I had never taught. Not once. Not a single student. In fifteen years.

Responsible citizens need significant, complex mathematical knowledge to actively participate in our democracy. Critically reading the news requires a conceptual understanding of large numbers, averages, distributions, statistical sampling, and voting systems. Where were these topics covered in the curriculum? Our courses that funnel students inexorably toward calculus never take the time to explore these "detours". And while some of my Survey students enjoyed the scenic overlooks of math-land, the Obama-McCain exchange made me aware of destinations much more important for them.

In the aftermath of my realization, I designed *Math for Social Justice*, a terminal math class for non-STEM majors, around a single question:

What mathematics do I want the person next to me in the voting booth to understand?

In addition to the topics above, we dove into estimation, tax rates, correlations, exponential growth, and probability. We read the news, practicing the skill of asking insightful questions about the numbers, statistics, graphs, and numerical arguments. To create a sense of ownership of at least some piece of mathematics, the students proposed, voted on, and carried out projects with the express purpose of improving the world.

These topics fit together to form a coherent course, one thoroughly enjoyed by both myself and the students, although many of the modules in this volume would have improved it. Rather than look at the specific content, the rest of this essay will delve into a few key characteristics of the course, some planned from the beginning and others the result of changes precipitated by early failures.

Leverage Their Interests. If you've taught a Math for Liberal Arts class, you probably noticed what I did: the students largely didn't like mathematics. Rather than dwell on this character flaw, the course tries to leverage what does compel them: politics, economics, the environment, and fairness. Concentrating on some local issue of interest, early in the semester each student had one minute in front of the class to propose a project to address that issue. The students voted, and, in groups, they attacked the most popular ones. Each proposal was required to do just two things:

- involve mathematics in some important way,

- improve the world (at least on the local level).

Shane proposed to dip into student fees to put solar water heaters on one of the dorms. Olivia pushed to change the cafeteria's to-go boxes to a more sustainable, biodegradable material. Peter, who has used a wheelchair since birth and requires live-in assistance, pitched the construction of campus housing with space for a caretaker; he wanted future students with disabilities to have the on-campus living experience he had been denied.

Only a fourth of the proposals were chosen by the class, and just a minority of those actually accomplished their goals, but that wasn't the point. When Shane's group calculated the payback time of solar panels, intersecting two linear functions, they sprinted down the hall shouting to anyone and everyone, "it's paid off in four and a half years!" They had used math to better understand an issue they cared deeply about. For them, mathematics had been transformed from a chore to being a useful tool, a lesson they might take with them as they work, live, and vote.

Use Current News. In 2009 when the economy hit rock bottom, our class dove deeper into probability and risk, eventually discussing how Wall Street used Credit Default Swaps and Collateralized Debt Obligations to bet on the housing market. A year later when the country was absorbed in the health care debate, we wove it into a discussion of percentages (drug effectiveness), distributions (health care spending is more concentrated among 30-somethings, more evenly distributed among 60-somethings), and exponential growth (7% growth per year is both exponential and unsustainable). When Mitt Romney released his tax returns, we spent a week examining how our income tax works, how the top rate has plummeted since the 1960s, and why Romney's overall rate could be so low.

Connecting theoretical topics to the news was enormously motivating for the largely math-phobic students. For a more STEM-loving crowd, the abstract idea that $\frac{dx}{dt} = 0.07x$ implies a doubling time of 10 units of time might be inherently compelling. For students in *Survey of Mathematics* — and for citizens — it's more important to translate the abstract math into everyday language and show them, or better yet, get them to discover for themselves, the real-world implications. Unchecked 7% growth will bankrupt Medicare, it's a mathematical certainty. What does a doubling time really mean? Even if you could somehow cut health care spending in half, but didn't change the 7% growth rate, you'd only extend the life of Medicare by about 10 years.

Teaching with one eye trained on the news served two purposes. For the students already following current events, this focus leveraged their interests to help them better understand the math. For the rest of the class, the discussions provided both peer and instructor pressure to care about current events, an important aspect of responsible citizenship.

Focus on Reasoning and Communication, not Computation. Future scientists certainly need to know how to take data and compute various statistics, using appropriate technological tools. What about (future) citizens? First, they should be able to interpret statistical evidence and communicate it effectively; future journalists are more likely to be sitting in our *Survey* classes than in *Vector Calculus*. You might ask students to do this activity:

> Given data on the U.S. prison population disaggregated by race and crime, write an article summarizing the data, including appropriate graphs and additional information as needed.

But I want more than that from the voter next to me. Critical thinkers know that every statistic presented in the media comes with a particular slant. How do we get students to understand the skewed nature of modern reporting? We add a twist:

> Take the same data and write two articles, one from the perspective of a "tough-on-crime" politician who sees the current system as justified, and the other from the perspective of a #BlackLivesMatter activist who sees racial bias in the data.

If we choose the data well, we can even end the activity by reading actual press releases from different groups.

Asking students to consider multiple perspectives helps them with the even trickier task they'll face when reading the news: given the perspective of a single article, imagine what an opposing view of the same issue (and same data) might look like.

Skills and the Inclination to Use Them. After an early version of *Math for Social Justice*, student performance on the final exam suggested that they had significantly improved the skills needed to critically examine quantitative claims in the media. However, student comments painted a more sobering picture: they didn't really plan to put those skills to use. Depressed, I looked back on the student progress like a grove of soundlessly falling trees, with nobody there to hear them.

In the next iteration, I added a new goal: develop the inclination to critically examine quantitative claims in the media. To do this, I launched a two-pronged attack.

First, I put more emphasis on actually reading the news, with daily discussion board assignments. Before each class, half the students had to post an article that made a quantitative argument; the other half had to ask an appropriate mathematical question about the numbers. (The next day the two groups switched roles.) For the first half of the semester, class discussion slowly built up students' understanding of what made a question good. For the last half of the semester, the first group posted both the article and the question, and the second group had to attempt an answer.

By practicing the art of asking good questions, answering them, and brainstorming as a class how to make both the questions and answers better, students developed both their skills and their inclination to use those skills.

I aimed to fuel that inclination even more with a second important change. To increase their motivation to build and use their (new-found) critical thinking skills, I carefully reframed the entire course. On the first day of class, I crudely pitched the course as teaching them the math they needed "so you don't get screwed over — and I mean 'you' both in the singular and the plural".

For some students, this was the hook they needed, the motif playing in the background as we read articles about the usurious payday lending industry that feeds off the desperation of the poor and tax policies that mostly benefit the 1% (or even 0.1%). We tore apart press releases from the federal government and from our own college public relations office, with critical thinking skills slowly replacing their initial naïveté. They dug through the details of candidates' tax plans, showing how some tax policies mostly benefited the 1% (or even 0.1%) — contrary to the rhetoric of those same candidates.

Nobody wants to be manipulated, and I regularly hear from former students who remember the course — and the content — because it was framed as a way to protect them from forces trying to take advantage of them.

Conclusion. Four years before the Obama/McCain debate prompted me to change my teaching, President Bush defeated Senator Kerry in a close vote. Part of the pre-election media narrative focused on which candidate people would rather have beer with. One poll of undecided voters found that 57% of them preferred Bush as a drinking partner (though for him it would have been a non-alcoholic one).

At the time, hearing that some voters might pull a lever based on something as frivolous as drinking-partner-desirability just angered me. Four years later, watching different candidates discuss small business profits, I came to a different understanding.

A well-meaning citizen without the quantitative reasoning skills to critically examine the content of the debate would find other ways to evaluate the candidates. Do the candidates' statistics sound like they can't possibly both be true? One of them must be lying. Without the mathematical education to decipher the arguments, you would naturally fall back on other ways to make choices.

The mathematics community has not only the opportunity to improve the mathematical knowledge of our students, but also the responsibility to prepare them to be citizens. In the process of designing a course around the core idea of responsible citizenship, I realized both the breadth and depth of the mathematics that is needed to be an informed participant in a democracy. The Math for Social Justice movement, as embodied in my course and the many important contributions contained in this volume, shows how we can rise to this challenge, toward a world in which all citizens have the quantitative reasoning tools to critically examine issues that affect us all, actively participating in our vibrant democracy.

Part C

Modules

7

Sea Level Change and Function Composition

Dawn Archey

Abstract. Although most carbon dioxide emissions have been produced by wealthier industrial societies, it is poorer nations that will be impacted first and worst by any resulting change in climate. One example of this is the dramatic impact sea level rise is expected to have on subsistence fisher-folk living on low-lying islands. In this module, groups of students work with a linear function and a quadratic function representing the sea level rise in the Republic of Maldives and create a simplified model of the geography of the Maldives (a triangular prism) to study the effects of sea level rise in the region. After they compute the geographical effects of sea level rise, they are asked to research and write about how this impacts livelihoods and society in the Maldives and similar island nations. Students will need to use and understand function composition, slopes, unit conversions, and similar triangles. The module is designed for a pre-calculus/college algebra course, but variations for calculus students are suggested. **Keywords:** Climate change, sea level rise, function composition, similar triangles, group work, writing.

1 Mathematical Content

This module is designed for use in pre-calculus. At my institution, this course begins with an introduction to functions, progresses through a variety of function types, and terminates with trigonometry. The module does not explicitly use any trigonometry, although it does use similar triangles. Thus it could be used in a college algebra course or a pre-calculus course which does not include trigonometry. However, it is particularly well suited to a course that involves trigonometry because it provides a chance early in the semester to bring in similar triangles, which will be needed later.

Before working on this module, students need to know how to compose functions and how to express a function as a composition of functions. They also need to be very familiar with lines and be able to compute with a quadratic function (composition and evaluation at a point). Finally, students need to know (or be reminded) that the sides of similar triangles have the same ratios.

This module seeks to address two common issues in pre-calculus courses. The first issue is that students often don't understand why we make them compose functions; they feel it is a meaningless exercise in symbol pushing. Hopefully, by composing functions with physical meanings (and dire consequences), the students will gain an appreciation of the power of function composition. This is one of the reasons why they are asked to work with two possible sea level rise functions. They must construct a function for area remaining in the Republic of Maldives depending on a function $r(t)$ for sea level rise before knowing the formula for $r(t)$. In this way, when the function changes (a different prediction is used) the students only need to make one substitution and do not need to repeat all the work.

The second issue addressed by the module is that many students have forgotten how to use similar triangles. This is problematic because when the trigonometry unit starts in a typical pre-calculus course, similar triangles are ubiquitous and often implicit. Even the definitions of the trigonometric functions require that similar triangles have the same ratios of side lengths. By using similar triangles in this module early in the course, students will be more familiar with them when they reappear in trigonometry.

2 Context / Background

Since the beginning of the industrial age, the concentration of greenhouse gases in the Earth's atmosphere has been increasing [19]. The most common greenhouse gas is carbon dioxide (CO_2). Greenhouse gases warm the earth by preventing some of the sun's energy from escaping back to space. The primary increases in concentrations of greenhouse gases in the atmosphere are from the burning of fossil fuels, transport, agriculture, and changes in land usage such as deforestation [6, pp. 45-56]. As Walker and King note in [19, p. 188], "The world's major industrialized countries ... have been responsible for almost all of the current climate problem; they gained their wealth and advanced state of development largely by exploiting cheap fossil fuels at an early stage". However, it is not these wealthier industrialized countries which will experience the most severe effects of the climate changes brought about by the increased concentrations of CO_2, but rather poorer nations such as those which are heavily dependent on agriculture, tourism, and fishing [3, 6, 7, 16].

Scientists believe that the global average temperature will increase by about 2.5° C by 2100 [16] as a result of increasing concentrations of greenhouse gases in the atmosphere [19]. A change of 2.5°C may not sound like much, but current global average temperatures are only 3 to 5° C higher than the last ice age [19] during which sea levels were about 122 m lower than they are today [17]. One of the consequences of the increase in temperature is rising sea levels. There are two mechanisms causing the sea level to rise: thermal expansion and ice melt [8]. If the ice sheets of Greenland and Antarctica continue to melt at their current steady pace, a modest sea level rise of 0.18 m will occur by 2100 [6]. However, many scientists believe the rate to be accelerating [9,17]. The Intergovernmental Panel on Climate Change (IPCC) predicts between 0.28 and 0.98 m of sea level rise by 2100 [4]. Some reputable sources predict up to 2 m of sea level rise by 2100 [6].

The human impact of rising sea levels is not uniformly distributed [19]. As Broecker and Kunzig observed in [3, pp. 142-143], "The worst impacts of the sea level rise would, of course, be felt in the poorer and even lower-lying countries, such as Bangladesh, where 35 million people live on the coastal floodplain, or Vietnam, or the various small island nations of the Pacific, some of which, such as Tuvalu, are already sinking beneath the waves". Rising sea levels will affect many aspects of life for people living in poverty and in coastal areas. Some of these aspects are: erosion [6], saline intrusion into fresh (drinking or agricultural) water [6], loss of ancestral home lands [6], loss of revenue from tourism and fishing [11], loss of revenue and utility of ports [6], increased incidences of mental health issues [6], coastal flooding [3,19], increased disease incidence (especially due to water-born pathogens with increased flooding) [3,16], social and political upheaval [6], and decreased crop productivity [6,19].

Some students may feel that the affected areas are remote, and so it is worthwhile to note that sea level rise is already affecting the coastline of the United States, with significant impacts projected later in the century for major cities such as New York and Miami. For perspective, it can be helpful to compare the population of these metropolitan areas with that of Bangladesh, one of the densest countries on Earth, to drive home the significance of impact abroad. In this module, I have chosen to focus on the Maldives because it was the lowest-lying nation for which sufficient data were available.

3 Instructor Preparation

There are two types of problems that are non-standard, or at least non-central, that the students need to be confronted with before working on the module. The first is a problem that asks about the meaning of various features of a linear function. I usually give a story problem and then require the students to explain the meaning of the y-intercept and slope in two complete sentences with units. Some students find it helpful to be told that units on the slope are always the units for the y-variable per the units for the x-variable. An example of such a story problem can be found on my blog, [1].

The second type of problem students need to have seen is an applied problem where they are asked to use function composition. The standard version is something like, "Suppose the radius of a circle is increasing at a rate of 3 cm/min. Write a function for the radius of the circle at time t, assuming the circle started with radius 1 cm. Now use function composition to write a function for the area of the circle at time t". This may be a good first example, but seeing a second more concrete example will help the students with the module.

The module contains a preparatory assignment, which will need to be distributed one or two classes before the day of the module. The instructor must do something to ensure the students do the assignment. For example, I have habituated my students to do such things by having "class prep" assignments for a grade every day.

The main part of the module is a worksheet for the students to work on in groups during class. As they work through the worksheet, the students will create and use two different functions for the area remaining in the Maldives. The first function will be linear and predict that 16.2% of the Maldives will have disappeared by 2100. The second function will be quadratic and predict that 38.4% of the Maldives will disappear by 2100. Both functions are consistent with historical data and predicted future sea level rise.

The worksheet takes longer than a 50-minute class period, but my students were able to reach problem 10 during class and the problems following that one are similar to problems encountered earlier in the worksheet. My usual solution for worksheets which take longer than a class period is to make extra copies of the worksheet. In the beginning of class, only one worksheet is given to each group. At the end of class, the group copy is collected and each student is given a fresh copy to finish on their own at home. This prevents one student from getting stuck with having to finish for everybody. Interested readers may contact me for the TeX source code for this module (which includes all answers as comments) so that they can modify it to fit their needs.

Also before class, the instructor must decide what kind of follow-up activity to assign. I suggest two follow-up writing assignments, designed to help the students reflect on the social justice aspects of the assignment. The high-intensity version has each group of students collaborating to write a short paper. The low-intensity version simply has each student write one paragraph. The high-intensity version will be more work for the instructor, even though the volume of text created will not be that much greater, because the students will need more coaching and the grading will require fact checking.

Additionally, either before the day of the module or during the module, the instructor should be prepared to discuss mathematical modeling briefly. Students may be unaware of the concept of making simplifying assumptions in order to make predictions and the idea that a line or curve fit to data may be very useful without being a perfect fit. The instructor may also need to make students aware, if the students feel the elevation is changing slowly, that the Maldives will likely become uninhabitable well before it disappears completely due to just being too small, groundwater salination, and/or high tides.

Some students may doubt that climate change is occurring or that climate change is being caused by humans. My response to such students would be two-fold. First, I would let such students know that although the news and politicians often speak as if the science behind climate change is inconclusive, the scientific community is actually in very strong agreement on this issue. Most of the leading scientific organizations worldwide have endorsed the position that climate change is due to human activity and 97% or more of actively publishing climate scientists also hold this position [15]. Second, I would point out to any doubting students that, even if there was any doubt, many people in the world cannot afford to wait [7]. This is the purpose of the video students are asked to watch at the beginning of the assignment. In addition, to remind students of the human side of the problem, on an exam a few weeks after the module, the students are asked how the people of the Maldives would respond to the character in the comic strip [14] who claims he will just wait and see on climate change to avoid the possible embarrassment of being wrong. In my experience students uniformly respond with strongly-worded statements against the comic strip character's position.

4 The Module

The module consists of a brief preparatory assignment to pique the students' interest, an in-class worksheet to be done in groups, and a writing assignment to be done after class. The in-class worksheet typically takes longer than a 50-minute class period; ways to handle this have been proposed in the previous section.

4.1 Preparatory Assignment. Before the in-class portion of this module, I assign my students some preparatory work to be done at home the night before. The handout I give them for this preparatory assignment is included as Appendix B.1.

I first ask them to watch the four-minute video available at:

`https://www.telegraph.co.uk/news/earth/earthvideo/6777182/The-sea`
`-is-killing-our-island-paradise.html`.

This video depicts the struggles of the people of the Carteret Islands as sea levels rise and make their homeland uninhabitable. I ask a few basic questions about the video just to be sure they watch the video. For example, I ask for their first impressions of the video and to name two factors making relocation difficult.

Regarding Questions 1 through 3, students' first impressions are typically about the sadness of the situation. The Carteret Islands are becoming uninhabitable due to consequences of global warming and sea level rise such as shortage of food and soil erosion. Unfortunately, it is no easy thing to move to the mainland of their nation to escape the rising sea level, because the mainlanders are not welcoming and don't wish to give up land for the newcomers. Additionally, many residents of the Carteret Islands do not want to leave their homes, and the mainland government has not properly allocated the promised funds for relocation.

Then, students are told in the handout that the sea level is rising due to thermal expansion and the melting of ice sheets in Greenland and Antarctica [8] due to global warming. The students are told the range of predictions for sea level rise between now and 2100 and asked to convert this information to millimeters per year in Question 4.

Finally, the students are introduced to the model of the Maldives we will use in class, namely a triangular prism. The students are asked to convert all units to kilometers and label the diagram with the measurements in kilometers in Question 5. Then in Question 6 students must compare the true area and coastline lengths of the Maldives to those in the triangular prism model as a way to verify the reasonableness of the model.

4.2 In-Class Worksheet. The in-class portion of the module involves a worksheet that students work together on in groups, a copy is included in Appendix B.2. In addition to the worksheet I usually also provide the students with the references.

In Question number 1, students are asked to draw the island model as the water is rising. Given the suggested labels for the width of the base of the (new) prism ($B(x)$) and the elevation of the new prism (x), students might end up with drawings that look like:

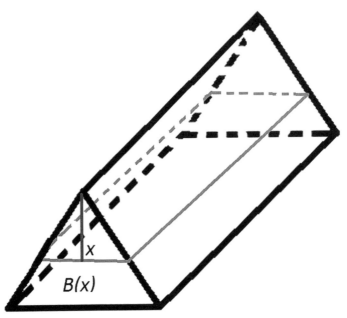

Next, in Questions 2 and 3, students use similar triangles to write formulas in terms of x (the elevation) for the base $B(x)$ of the prism and the "map area" $A(x)$ of the prism. The map area in this project means the area you would see by looking down on the prism from above (i.e., the area it would take up on a map). In Question 4 the students use subtraction (and unit conversion) to relate the amount the sea level has risen, z to the current elevation $E(z)$. Later, when the function composition begins x will be replaced by $E(z)$.

Questions 5 through 7 deal with the meaning of function composition. The table in Question 5 is designed to help students see how the output of one function becomes the input of the next in function composition. When students ask for help with this question I usually end up drawing a sort of zig/zag or Z pattern in the table which helps them see the meaning I am trying to convey.

At the end of Question 7 the students will have a function $A(t)$ which gives the map area of the model of the Maldives in year t, but it is not an explicit function. It is really a function in terms of $r(t)$, where $r(t)$ is the sea level in the Maldives t years after 2010. In Questions 8 through 11 the students work with one possible function for sea level, namely a linear function. Then in Questions 12 through 14, the students work with a quadratic formula for sea level. In both cases the students produce an explicit function for $A(t)$ and use it to predict the remaining area of the Maldives in 2100.

There are two reasons why I introduce the two possible formulas for $r(t)$, a pedagogical reason and a climatological reason. The pedagogical reason is highlighted in Question 13, namely that because we used function composition we did not have to start all over when the formula for sea level changed. The climatological reason is that as shown in [9], it is likely to take another century before we are able to determine whether sea levels are rising in a linear manner or in an accelerating manner such as a quadratic formula would indicate. When I mention this finding to students they are generally surprised and also feel better about me forcing them to "do the same thing twice".

4.3 Writing Assignment (after class). In Appendix B.3, two possible writing assignments that can wrap up the module are described. The goal of these writing projects is to make sure students think about what they have done and reflect on the human implications of sea level rise.

Each assignment asks students to choose one specific consequence of sea level rise (for example loss of revenue from tourism) and write about its impact on the people of the Maldives. One is a high-intensity writing project for groups resulting in a two to three page paper. The other is an individual writing assignment which involves some research leading up to a single paragraph. However, you can choose a different type of writing assignment that seems to fit your situation. For example, one year I had my students prepare posters for a college-wide event showcasing undergraduate research.

Through these writing-intensive wrap-up activities, students get to explore the social impact of the issue they have explored mathematically earlier. Besides reflecting on the social justice implications of sea level rise, students also have the chance to practice writing about scientific topics in a non-technical way, which is an important job skill.

5 Additional Thoughts

5.1 Application Notes. I have used this module in my pre-calculus class. I assigned a variation on the high-intensity writing project. Instead of writing it up as a paper, the students were required to produce a poster to present at a student research symposium held at my institution for alumni and local high school students. With help from their instructor, the students produced posters which interested many symposium attendees. The students found that the opportunity to talk about their work was enjoyable and helped them synthesize what they had done. I can email a copy of the posters upon request.

The problems the students had the most difficulty with were writing the functions for elevation and area. There were two separate issues here. One is that students had trouble internalizing that the elevation and area were changing. They would make statements like "But we know what the area is; it is 320". To help them visualize this, I have made a physical model out of plastic canvas and yarn, although a similar model could be made of cardboard and tape; see Figure 7.1.

The second issue is that students had forgotten that similar triangles have the same side ratios. Once reminded of this fact by a drawing on the board they could proceed.

Although this activity challenged students, they liked it. They reported that it helped them see the real-world applicability of mathematics and brought their attention to a significant issue that they rarely hear about. One student said that even though he recognized the unit conversions were easy, he found them challenging and thought that was good practice. Another student made my day by saying "The writing was the hardest part because we had to interpret and say in words what we did".

Figure 7.1. A physical model made out of plastic canvas and yarn.

5.2 Possible Variants. One possible modification is to allow the students to design their own model shape for the islands. They might design a shape that does not require similar triangles, if given complete free rein. However, a cone or several cones would also require similar triangles. Students could evaluate which of several possibilities seems the most realistic.

This activity could also be modified for use in other courses. A sea level activity for differential calculus could be created by using similar functions and employing the chain rule. Alternately, one could construct a related rates problem. For example, I provided my calculus students with only a little of the background information, asked them to do Question 5, and then asked them "At what rate is the area changing when the elevation is 2 meters if the sea level is rising at a rate of 4.3325 mm/year?" and "The Intergovernmental Panel on Climate Change (IPCC) predicts between 0.28 and 0.98 m of sea level rise by 2100 [**4**]. Some reputable sources predict up to 2 m of sea level rise by 2100 [**6**]. What is the annual rate of sea level rise in mm per year for each of these estimates? What is the rate of change of the area for each of these estimates?" The full text of the related rates version I created of this problem is available on my blog [**2**]. The related rates version only takes about half an hour as compared to over 50 minutes for the precalculus version.

For integral calculus or multivariate calculus, one could replace the model of The Maldives as a triangular prism with an exercise using Riemann sums and a topographical map. Those who wish to use calculus to model the average rise of sea level worldwide may be interested in "Mathematical Models for Global Mean Sea Level Rise" [**13**], which focuses on estimates attributed to the two primary drivers of increase, thermal expansion and ice-melt.

Acknowledgments. The author would like to thank Mark Benvenuto, Vesta Coufal, Sarah Isaksen, and Shelby Long for their useful suggestions.

Appendix A: Tables and Datasets

The annual average sea level data from Hulhule Island, the Maldives 1989 to 2013 from
[18] is given in the table below.

Year	Sea Level (mm above 2010 level)	Year	Sea Level (mm above 2010 level)
1989	-129.73	2001	-69.83
1989	-129.73	2002	-48.15
1990	-107.62	2003	-39.95
1991	-65.01	2004	-32.08
1992	-68.88	2005	-63.93
1993	-75.09	2006	-26.51
1994	-34.22	2007	-17.31
1995	-46.14	2008	29.51
1996	-54.87	2009	-8.32
1997	-27.95	2010	0
1998	-37.58	2011	20.50
1999	-34.46	2012	32.03
2000	-48.56	2013	31.01

The line $y = 4.6901x + 5.2854$ is the best fit line, with $R^2 = 0.7287$ for y being
sea level in mm above the 2010 level and x being years since 2010. However, if we ask
instead for the best fitting line which agrees exactly with the 2010 sea level (i.e., goes
through the origin), we obtain the line $y = 4.3325x$ with $R^2 = 0.7218$ as shown in the
graph below. The data is very noisy due to annual variations from events such as El
Niño, so this value of R^2 is reasonable.

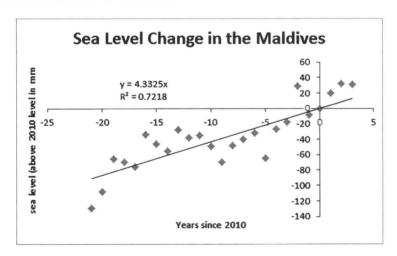

Appendix B: Assignments and Handouts

B.1 Preparatory Assignment. During the next class, we are going to use function composition to investigate the effects of sea level rise. We will focus our attention

on the Republic of Maldives, but there are many other low-lying islands facing a similar plight. Here is some material to prepare you:

(1) Watch this video (4 minutes) which provides context and human motivation for our work in the next class:

https://www.telegraph.co.uk/news/earth/earthvideo/6777182/The-sea
-is-killing-our-island-paradise.html

What are your first impressions?

(2) What is causing the Carteret Islands to become uninhabitable?

(3) Name two factors making relocation difficult.

****

Sea level is rising due to thermal expansion and the melting of ice sheets in Greenland and Antarctica [8]. The Intergovernmental Panel on Climate Change (IPCC) predicts between 0.28 and 0.98 m of sea level rise by 2100 [4]. Some reputable sources predict up to 2 m of sea level rise by 2100 [6].

****

(4) If sea level rises at the lower level predicted by the IPCC, how many mm does it rise per year?

****

The reason the sea level is rising is due to rising global average temperatures. Scientists now believe that the global average temperature will increase by about 2.5° by 2100 [16] as a result of increasing concentrations of CO_2 in the atmosphere [19]. This may not sound like much, but current global average temperatures are only 3 to 5° C higher than the last ice age [19], during which sea levels were about 122 m lower than they are today [17]. The increased levels of CO_2 are mostly due to the activities of wealthier industrialized nations [19]. For example, in 2010, the USA's per capita CO_2 emissions were 17.6 metric tons, but the Maldives per capita CO_2 emissions were 3.3 metric tons [20]. Unfortunately, despite not being a major contributor to the problem, The Maldives is an example of a poor low-lying country which will be hit soon and hard by the effects of global climate change including sea level rise [6], [1]. The Maldives is located south-southwest of India. It consists of approximately 1,190 islands and is the lowest country in the world with a maximum elevation of 2.4 m. The Maldives has a total area of 298 km squared and a coastline of 644 km [1].

****

(5) Since the actual shape of the Maldives is complicated, we will use a simpler shape to compute with and assume this is a good stand-in for the real islands. We will pretend the nation is one triangular prism as pictured below. The width of the base is 1 km and the length of the base is 320 km. The height of the prism is 2.4 m. Convert all units to kilometers and label the picture with measurements in kilometers. What is the map area of the prism in km^2? By map area, in this project, we mean the area of the rectangle you would see by looking down on the prism from directly above it.

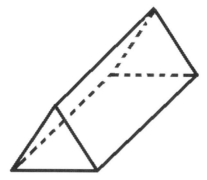

(6) How close to the true area and coastline measurements is this approximation?

****
It could have been closer, but for simplicity, the lengths were rounded to the near-est kilometer. You might be interested to know that using calculus one can show the average height of the prism is 1.2 m, which is a little lower than the average height of the true Maldives which is approximately 1.5 m [12].
****

B.2 In-Class Worksheet.

(1) As the sea level rises, the prism gradually becomes covered with water. The island above water is still prism shaped. Draw what this looks like. Label the width of the base of the (new) prism $B(x)$ and the elevation of the new prism x.

(2) Give a function $B(x)$ for the width of the base of the prism in km when the elevation of the prism is x km. Hint: Use similar triangles.

(3) The length of the base of the prism never changes (because it has sheer cliffs on the ends). Give a function $A(x)$ for the map area of the prism in km^2 when the elevation of the prism is x km. Recall that by map area in this project, we mean the area of the rectangle you would see by looking down on the prism from above.

(4) Give a function $E(z)$ for the elevation of the Maldives in km when the sea level is z mm above the 2010 level.

****
Let $r(t)$ be a formula giving the sea level in the Maldives t years after 2010 in mm above the 2010 level. We will not give a specific formula for $r(t)$ at this time because we want to study the effect of different patterns of sea level rise on the Maldives.
****

(5) Fill in the following table:

Function	inputs (with units)	outputs (with units)
$A(x)$		
$E(z)$		
$r(t)$		
$E \circ r(t)$		
$A \circ E \circ r(t)$		

(6) What is the meaning of the function $E \circ r(t)$?

(7) Use function composition to write a function $A(t)$ which gives the area of (the prism model of) the Maldives t years after 2010 in km^2.

****

Suppose the linear function $r(t) = 4.3325t$ is used to predict the sea level in the Maldives t years after 2010 in mm above the 2010 level. This function is consistent with the historical data as shown in the appendix.
****

(8) The function $r(t)$ is a linear function. What are the units of the slope and y-intercept of $r(t)$? What are the significance of the slope and y-intercept of $r(t)$?

(9) Using $r(t)$ as given above, give an explicit formula for $A(t)$.

(10) According to the prism model, what will be the remaining land area in km^2 of the Maldives in 2100? What percent of the Maldives will have disappeared in 2100?

(11) Where does the formula for $r(t)$ fit in the range of sea level rise predicted by the IPCC? Are the predictions we just made on the optimistic end or the pessimistic end?

****

A linear formula for sea level rise, such as $r(t)$, assumes that the ice sheets of Greenland and Antarctica [6] continue melting steadily at their current rate. However, increasingly scientists believe the melting rate is accelerating [9], [4]. Now suppose instead that the quadratic function $r(t) = 0.068t^2 + 4.1t$ is used to predict the sea level in the Maldives t years after 2010 in mm above the 2010 level. This function is consistent with both historical data and the prediction of about 1 m of sea level rise over the next hundred years [9], [10].
****

(12) Use function composition and your previous work to give a formula for map area remaining in km^2 in the prism model of the Maldives after t years if the sea level rises according to the quadratic function $r(t)$.

(13) Was your work in Problem 12 made easier because your area function in Problem 7 was given in terms of a generic function giving sea level rise? If so, how?

(14) Use your function from Problem 12 to determine what will be the remaining land area in km^2 of the Maldives in 2100. What percent of the Maldives will have disappeared in 2100?

****

Although some further warming is inevitable due to the effects of carbon already in the atmosphere [8], [19], much of the technology needed to halt global warming and its effects such as sea level rise already exists (or is very nearly developed) [19]. However, there is much debate over who should pay to implement such carbon reducing measures and whether it is better to implement them now when less damage has been done or in the future when it may be cheaper to implement because of technological advancements [19]. Additionally, there is also great debate about who will pay for such measures [19]. Many scientists also fear that a time is rapidly approaching when even if we halted carbon emissions completely, some effects would be irreversible [17]. For example, once an ice sheet breaks off, it is not cold enough to re-form.

****

B.3 Writing Assignment (after class).

High-Intensity Writing Assignment. As a group you will write a paper about the impact on the people of the Maldives from one consequence of sea level rise. Your group must choose one particular consequence of sea level rise to focus on from the list below. Cite your sources.

- Student 1: Write a paragraph of introduction and a paragraph of explanation of the mathematics from the worksheet.

- Student 2: Write a paragraph on the science/logic/meaning behind your chosen consequence. For example, if you choose ground water salinization explain the physical mechanisms by which this occurs. As another example, if you choose loss of ancestral homeland, use anthropology to describe why ancestral homelands are important. Also find some pictures or graphics to illustrate your group's paper.

- Student 3: Write two paragraphs explaining how this consequence affects people in general and the Maldives in particular.

- Student 4: Write a paragraph of conclusion and assimilate all the other paragraphs into a paper. Make sure there are transitions and the paragraphs are arranged logically.

Possible consequences to focus on: erosion [6], saline intrusion into fresh (drinking or agricultural) water [6], loss of ancestral homelands [6], loss of revenue from tourism and fishing [11], loss of revenue and utility of ports [6], increased incidences of mental health issues [6], coastal flooding [3], [19], increased disease incidence (especially water-born pathogens with increased flooding) [3], [16], or social and political upheaval [6].

Low-Intensity Writing Assignment. Do some outside research and write a paragraph about the impact on the people of the Maldives from one of the consequences of sea level rising listed above. Cite your sources.

Bibliography

[1] D. Archey. "Global Temperature–Meaning of Slope and y-intercept." Web blog post. *Dr. Dawn's Blog.* Full text at `http://blogs.udmercy.edu/archeyde/2015/07/07/global-temperature-meaning-of-slope-and-y-intercept/`. Accessed on May 19, 2016.

[2] D. Archey. "Sea Level Rise Related Rates Problem." Web blog post. *Dr. Dawn's Blog.* Full text at `http://blogs.udmercy.edu/archeyde/2016/05/18/sea-level-rise-related-rates-problem/`. Accessed on May 19, 2016.

[3] W.S. Broecker and R. Kunzig, *Fixing Climate,* Hill and Wang, New York, 2008.

[4] J. A. Church, et. al. "Sea Level Change". In: *Climate Change 2013: The Physical Science Basis. Contribution of Working Group I to the Fifth Assessment Report of the Intergovernmental Panel on Climate Change.* Cambridge University Press, Cambridge, United Kingdom and New York, NY, USA. 2013. Full text available at `http://www.climatechange2013.org/images/report/WG1AR5_Chapter13_FINAL.pdf`

[5] The CIA World Factbook, `https://www.cia.gov/library/publications/the-world-factbook/geos/mv.html`, accessed on May 6, 2016.

[6] K. Dow and T. E. Downing, *The Atlas of Climate Change: Mapping the world's greatest challenge,* 3rd ed. University of California Press, Berkeley, 2011.

[7] Francis, Encyclical Letter *Laudato si',* Vatican City, 18 June 2015. Full text at `http://w2.vatican.va/content/francesco/en/encyclicals/documents/papa-francesco_20150524_enciclica-laudato-si.html`, accessed on May 6, 2016.

[8] T. Flannery adapted by S. M. Walker, *We Are the Weather Makers: The history of climate change,* Candelwick Press, Massachusetts, 2009.

[9] I D. Haigh *et al.,* *Timescales for Detecting a Significant Acceleration in Sea Level Rise* Nature Communications 5, Article number: 3635. 14 April 2014.

[10] I. D. Haigh, personal correspondence, 2014.

[11] L. Hannah, *Climate Change Biology,* Elsevier, New York, 2011.

[12] J. Henley, "The Last Days of Paradise". *The Guardian* (*London*). 11 November 2008. Accessed 17 June 2015. Full text available at `http://www.theguardian.com/environment/2008/nov/11/climatechange-endangered-habitats-maldives`, accessed on May 6, 2016.

[13] Stephen Kaczkowski, *Mathematical models for global mean sea level rise,* College Math. J. **48** (2017), no. 3, 162–169, DOI 10.4169/college.math.j.48.3.162. MR3654841

[14] R. Munroe. "Playing Devil's Advocate to Win" `https://xkcd.com/164/`. Accessed June 17, 2015.

[15] NASA. "Consensus: 97% of climate scientists agree." Accessed 17 June 2015. Full text available at `http://climate.nasa.gov/scientific-consensus/`.

[16] W. D. Nordhaus and J. Boyer, *Warming the World: Economic Models of Global Warming,* MIT Press, Cambridge, MA. 2000.

[17] F. Pearce, *With Speed and Violence: Why scientists fear tipping points in climate change,* Beacon Press, Boston, 2007.

[18] University of Hawaii Sea Level Center. `http://uhslc.soest.hawaii.edu/data/download/fd` accessed July 23, 2014.

[19] G. Walker and D. King, *The Hot Topic: What we can do about global warming,* Harcourt, New York. 2008.

[20] The World Bank Data Bank, `http://databank.worldbank.org/data/home.aspx` accessed July 23, 2014.

8

Exploring the Problem of Human Trafficking

Julie Beier

Abstract. Despite progress made in human rights around the world, slavery and the exploitation of humans still persist. Human trafficking is a global market involving the transport and sale of adults and children for forced labor or sexual exploitation. In this module, we seek to better understand human trafficking networks by utilizing graph theory. First, students interact with one or more texts related to human trafficking in the current century, learning about the problem and hearing the stories of victims. Next, we look at the mathematics of introductory graph theory to model networks, assuming no particular mathematical background for students. Students look at a sample social network to further learn the basic measures of interest. Airline maps and highway maps are then used to explore the current human trafficking network in the United States, and students are asked to reason about the policy and legal implications of their findings. Finally, we discuss other connections possible in this unit including writing instruction, vocation exploration, and other network applications. At a minimum, students finish this unit with an awareness of the human trafficking problem, introductory knowledge in network theory, and, most importantly, an example of how problems in today's world can benefit from the application of mathematics. **Keywords:** Human trafficking, networks, graph theory, in-class discussion, team work, labs.

The goal of this module is to bring forth an understanding of human trafficking in the United States using tools of graph theory. To put this troubling social issue into a deeper context, students are provided readings and other background information, along with guidance for discussion. Building on this knowledge, students undertake the task of modeling networks, in an effort to understand how such models could aid law enforcement or inform public policy. An extensive version of this module can span over eleven class periods, but alternate versions are presented.

This module is also designed to help students understand the realities of modeling and data analysis. The network maps take time to complete and require teamwork, both important real world lessons. The mathematical model we use is imperfect and yet supports existing data; this should lead to interesting discussions with students.

1 Mathematical Content

This module could be utilized in any liberal arts mathematics course or a non-mathematics course with a focus on social justice issues. No mathematical background is assumed for students beyond basic numeracy skills. It should be noted that the instructor is also not expected to be an expert in graph theory or networks, but there is space for the more advanced graph theorist to expand the module.

In this unit students are introduced to the basic ideas of networks and graphs. This includes the definition of graph and network, vertex degrees, edge weighting, and minimal spanning trees. However, the focus is not on technical definitions and procedures. Rather, the focus is on expanding our knowledge of the issue of human trafficking using a tool not often brought to the problem: mathematics.

2 Context / Background

René Descartes said, "With me, everything turns into mathematics". If our view of mathematics is as wide as Descartes', then we see mathematics in the obvious places, like engineering and the sciences, but we also see it in the humanities and the social sciences. This motivated me to develop a course titled *Social Justice: Adding a Mathematical Dimension* at Earlham College, a small, national liberal arts school. This course is an option that students can select for their first-year seminar, is capped at a size of 16 students, and meets four days a week. Informed by the Quaker tradition, Earlham College documents its community values in *Principles and Practices* [5], a statement that defines and guides how we live and work together. In the course, this document is used to establish a common understanding of the meanings of social justice and a socially just society. The course then progresses through four units, each one focusing on a social justice issue and mathematical topic that is subsequently used to enhance understanding of the issue in some way. The module presented here is a unit from the first iteration of this course, and is presented in the format used for the course: a social justice topic, a mathematical topic, and the connection between the two. Note that this order is not necessary; it is possible, with minor modifications, to present some of the social justice and math content simultaneously. Since this module originates from a course for students in their first semester, no prior background, mathematical or otherwise, is expected at the start.

While slavery was formally abolished in the United States by the Thirteenth Amendment to the Constitution in 1865, that decision does not mark the end of the enslavement of people in this country, and certainly not in the world. Human trafficking is a global market involving the transport and sale of adults and children, typically for forced labor or sexual exploitation. The United Nations recognizes the problem of human trafficking under the Office of Drugs and Crime (UNODC), and posted in September 2014 on its website (`http://www.unodc.org`) that the topic of "human trafficking tops agenda at Transnational Organized Crime Conference". In order to address this "modern form of slavery", the UNODC established the Global Initiative to Fight Human Trafficking, and released a report in 2009 titled *Global Report on Trafficking in Persons*. Both a summary and the full report are available on the UNODC website; specifically, see [**15**].

The UNODC report utilizes data from 155 countries and presents some startling statistics. At the time of the report, only 63% of these countries had passed laws against the trafficking of people. Approximately 79% of human trafficking is for sexual exploitation, and approximately 79% of all victims are female. Forced labor accounts for about 18% of the reported trafficking. The percentage of humans trafficked that are children is 20% globally, although in some parts of the world it is as high as 100%. Only 30% of countries provided gender information about traffickers, but the reported data illustrates that women trafficking other women is surprisingly common.

The UNODC recognizes that not enough is known about the problem.[1] Governments sometimes obstruct the process of information gathering, there is a scarcity of data, and often there is neglect when it comes to reporting and prosecuting these cases. The maps accompanying the report illustrate that in some areas the transport of trafficked people is fairly regionalized, simply crossing into neighboring countries. However, over half of the victims in the United States come from other countries.

While the statistics in reports such as [**15**] are uncomfortable, they pale in comparison to the stories of the victims themselves. Victims often suffer abuse in a variety of forms, are arrested for the behavior forced by traffickers, and fear speaking out. This makes the stories of victims who choose to share even more important as we seek to fill in the gaps created by lack of data.

3 Instructor Preparation

3.1 Supporting Students.
The subject matter of human trafficking may be emotionally challenging for a number of students, particularly if students hear victims' stories. Thus it is important that a safe and respectful classroom environment be established before starting this module. There are many ways this can be implemented, including classroom guidelines for discussion, practice with less intense topics, and starting with silence to encourage students to bring their best self to a conversation. At Earlham, it is common for events to begin with a minute of silence in order to quiet ourselves, coming to the present moment, and we find students quickly resonate with this part of the Quaker tradition. It is important to explain the purpose of silence to

---

[1]A recurring justification for the failure to implement policies and procedures that aid victims is insufficient data. But the simple, accessible mathematics in this module provides a way to enrich the data and counter this argument. This impresses upon students that even basic tools from one discipline can solve problems in another.

students before implementing this, and to ensure the silence is long enough to allow everyone to settle.

It is recommended that, before beginning this topic, professors have information about counseling services available for students in case it is needed. They should also consider finding a technique to use in class that allows students to process emotionally intense moments before requiring conversation. For instance, after watching a survivor's story, students may be encouraged to make a list of their reactions or to respond to a specific portion of the story before discussion begins. Similarly, it may be useful for instructors to direct students' attention with focused questions before watching a story such as: "During the conversation, pay particular attention to how aid organizations can empower or continue to victimize. Does the organization discussed here empower or victimize? How do you know?" It is also necessary to divorce assessment from the emotional component. This is natural in the case of mathematical activities, but less clear in assignments that ask students to reflect or respond.

For conversation days, it is useful to have the room set up in a circle so that all students can see each other. An instructor who has not had discussion or seminar days previously in class should start by establishing clear guidelines for respectful communication, and practice having all students speak up using a favorite icebreaker activity. Students sometimes fail to learn each others' names; having students display their names during discussion might help.

Since the social justice portion of the module ends with aftercare, or how to support victims after they escape, and the mathematical portion ends with reflections on policy and law, students may be interested in doing something tangible after the unit. Instructors adopting this module can consider gathering the contact information for any local or regional organizations that support victims of human trafficking and asking about volunteer opportunities.

3.2 Materials. The required reading [7] is designed to introduce the subject of human trafficking. However, there are several supplementary materials that one may choose to use [3, 12, 14], and will need to gather before starting. All are currently published and should be available in a college library or through interlibrary loan. Instructors should preview the supplementary materials themselves to decide which ones to use. If they choose to show the DVD [3], they do need to remember to check the technology in their classroom.

Before beginning the mathematical activities, it is necessary to gather data. A quick search online will reveal a plethora of United States highway maps such as the one at [6]. Students will need to access this as well as airline route maps throughout the activity. If the class meets in a room with wireless Internet access and a sufficient number of students have laptops, or if the class meets in a lab, one will need nothing more than this link. However, if this is not the situation, the instructor will need to modify the lab to utilize printed maps and remove the built-in dependence on the Internet.

The mathematical activities are designed for teams of size three to four. It is important to discuss expectations for group work and cooperation before starting these labs. It would be useful to have tables or floor space for students to spread out on these class days as they will need to work with either a laptop or several map printouts while they are drawing graphs. While not strictly necessary, copies of a map of the United States would be useful.

Table 8.1. A timeline for the module.

Day	Prepare for Class	In Class
Day 1	Read [7] through P 64	Introduction to human trafficking
	Optional: Read [14] "Facts: Sex trafficking"	Class Discussion
Day 2	Read [7] through P 118	Class Discussion
	Short reflection on environment & identity	
Day 3	Read [7] through P 178	Class Discussion
Day 4	Finish [7]	Class Discussion
	Reflection on the role of community	
Optional	Optional: Read [14] "Facts: Aftercare"	Watch *Half the Sky* and discuss
Optional		In Class Debate (See Section 5)
Day 5		Introduction to graph theory
Day 6	Data Collection for Introductory Lab	Introductory Lab
Day 7	Read [14] "United States"	Introduce Human Trafficking Lab
		Start lab 2 (airline portion)
	Finish Introductory Lab	
Day 8	Finish through #2 on lab 2	Continue airline portion of lab
Day 9	Finish through #5 on lab 2	Start highway portion of lab
Day 10	Finish through #7 on lab 2	Continue with lab
Day 11	Finish lab reflection (#12)	Wrap up unit

4 The Module

In this section we present the module, including the necessary social justice and mathematical content. In order to aid in implementing the module, possible class prompts and questions are provided. These portions are imbedded in the text below. New definitions and terms for students are italicized. The questions and style of this section is meant to provide content that is directly usable for discussion and lecture. Sample handouts for the mentioned labs are provided in the appendix.

The module takes approximately eleven fifty-minute class periods, but can be shortened by reducing the readings and labs, or extended as discussed in the last section. Table 8.1 summarizes an eleven-day implementation of the basic module with a few optional components. A shortened version and other extensions are discussed in Section 5 of this chapter.

4.1 Human Trafficking. We begin by learning about human trafficking through texts. Students read *Sold* by Patricia McCormick [7] over the course of approximately four class periods. *Sold* is a work of fiction following the story of Lakshmi, a trafficked girl, that is based on stories of Nepali girls sold into sexual slavery in India. The fictional content allows some emotional distance when exploring this issue but its factual foundations allow it to be used as a reasonable starting point. Since this text is a young adult text, it is reasonable to expect students to read approximately 70 pages between class meetings.

In class, students can explore the text through discussion. Some possible questions to ask about this text include:

- *What surprises you about the process of trafficking women?*
- *Who has power in the situation?*
- *What power does the victim have, if any?*

- *What role does manipulation play in the story?*
- *Who is responsible for the fate of the women in the story (for example, is it the family, the girl, traffickers, customers, or society)?*
- *What are sources of hope for victims?*
- *How are victims controlled?*
- *How do victims support each other? Is this support a positive or does it only reinforce the institution?*
- *What are the economic, education, and health issues related to this situation?*
- *How does the problem of human trafficking affect those outside of the institution in society?*
- *How does the neglect of society lead to the victims in this story?*
- *What parts of society depend on the continuation of the enslavement of women?*
- *What are the turning points in Lakshmi's story? How do Lakshmi's decisions at these points in time change her situation?*
- *How does Lakshmi's internal self change during the story?*

During the first half of the unit, it is helpful to supplement the text with non-fiction texts. The text *God in a Brothel* by Daniel Walker [14] is the story of a man who combats the human trafficking institution, often by going undercover. Scattered in the book are pages that summarize facts related to the human trafficking industry that may be introduced during the class to contextualize and reinforce ideas. Sections of particular interest are "FACTS: Sex Trafficking", "FACTS: HIV/AIDS", and "FACTS: Aftercare". These naturally support the introduction, middle, and end of [7] as the main character, Lakshmi, encounters each of these. Students could inquire:

- *How does Sold resonate with these facts?*
- *Are there any points where they disagree?*
- *Are there things from Sold that you assumed were fiction that you now see are fact?*

Since *Sold* is set abroad, it may be easy for students to believe human trafficking and slavery are not problems in the United States. Chapter 11 in [14] shares the story of a trafficked woman in Atlanta, Georgia, and can be read to counteract this. This chapter also makes it clear that ending human trafficking requires the support of politicians and law enforcement. Possible discussion questions include:

- *What role does society play implicitly or explicitly in condoning human trafficking?*
- *How does the fact that Emily lives in Atlanta change how you hear her story?*
- *What do you think happens to Emily after this encounter?*
- *What bias is in this text, if any? How do you know?*

Note that [14] is written with a religious perspective that may influence which sections you choose to use in class.

Another possible supplement is a DVD titled *Half the Sky* [3], which is related to the book by the same name [2]. It focuses on social justice issues related to women around the world. The full video is long and would take several class periods to show, but this may not be necessary. The first five minutes establish the point of the movie and text, and provide context. The second section of the film, between approximately minutes 45 and 80 of the first disk, focuses on the story of survivors in Cambodia. During the episode, Somaly Mam, who was sold into the sex trade as a child, shows us her organization of young women who are survivors that support each other and help to educate the society. This gives the class an opportunity to discuss ideas such as:

- *How do you care for survivors?*
- *What actions empower women in this situation?*
- *How do these women change the society around them?*
- *How does working for change serve to heal?*

This section serves well to discuss aftercare and support of trafficked women. It is worth mentioning that the DVD also includes an interview with America Ferrera who works against forced prostitution in the Red Light District of India, which can be connected to a discussion of *Sold* if this is preferred.

4.2 Introduction to Graph Theory and Networks.

The mathematical tool that is used in this module to better understand human trafficking is graph theory. In particular, we let graphs represent networks. Instructors may want some resources to orient themselves to this particular area of graph theory. A free resource available online with low print cost is Maarten van Steen's *Graph Theory and Complex Networks* [13]. A sample text with coverage of a variety of applications is *Networks, Crowds and Markets: Reasoning about a Highly Connected World*, written by Easley and Kleinberg. [1]. The MAA textbook *Graph Theory: A Problem Oriented Approach* by Daniel Marcus [4] is a nice introductory text appropriate for undergraduates. In particular, instructors not familiar with graph theory may find the information about minimal trees in Chapter E of [4] particularly helpful. The minimum information that students need to know follows below.

In class the instructor can begin by activating students' previous knowledge by inquiring:

- *What does the word graph mean to you?*
- *What is a graph?*

It may be helpful to draw different pictures to help students see that mathematicians use graph to refer to graphs of functions and graphs as a set of vertices and edges. In this section, graphs are of the second type.

Below is a transcript of how I might go through this material in class with my students:

*A **graph**, G, is a collection of vertices, called V, and a collection of edges, labelled E. We write this graph as G(V, E). For example, G({a, b, c}, {ab, bc}) is the graph with vertices a, b, and c, and the two edges ab and bc. This graph looks like the left one in Figure 8.1. Does it matter where I put each vertex? What if I put b above a and c; is this the same graph? How do you know? Sometimes vertices in graphs are called nodes. An edge connects two vertices. We call the connected vertices **end points** of the edge. So ab has endpoints a and b. Two nodes or vertices are said to be **adjacent** or **neighbors** if there is an edge between them.*

Graph theory is interesting in and of its own right, but it also has many applications. Perhaps each of these nodes represents a person and there are edges between people who know each other, so person a and b know each other. Similarly, b and c know each other but a and c do not know each other. Now suppose the graph instead looked like the right graph in Figure 8.1. In this example, who is friends with f? What about b or j? The node j is an **isolated vertex** *and represents someone who does not have friends yet.*

Note: We could draw an edge from j to j, assuming that j is friends with him or herself. This would be a **loop**. *But since it is reasonable to assume that everyone is friends with themselves, we choose not to do this. When a graph can be separated completely into two subgraphs (such as in this example), it is* **not connected**. *Each piece that cannot be split is called a* **component** *of the graph.*

Now notice that stating who is friends with whom can take a long time if you have a large graph with lots of people and connections. One way to keep track of this information is to make an adjacency matrix, which has a row and a column corresponding to each node. The entry $a_{i,j}$ is 1 if there is an edge between i and j. Otherwise it is 0.

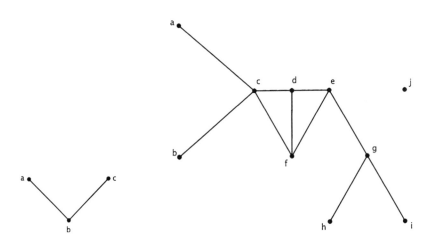

Figure 8.1. Example Graphs

I would then create the adjacency matrix with students for this example and continue with:

Which person in this example is the most popular? How do you know? The number of edges at a vertex is called its **degree**. *What is the degree of an isolated vertex? Still viewing the graph as a model of a friend* **network**, *who do you guess is likely to become friends next? Why?*

Here we want students to notice that pairs like f and g are more likely to become friends next than f and h or f and j since only a couple of edges are needed to go between f

and g, and there are multiple ways to connect them. This is a natural place to introduce the concept of tie strength if desired. The instructor might continue with:

> *Suppose that a wants to pass a note to h but each person will only pass the note to someone that is a friend. There are several ways that this could happen but we could choose: acdegi. There is an edge between each pair of vertices that are next to each other in this list. We call this a* **path**. *Another way to define connectivity of a graph is to say that a graph is connected if for every two vertices there exists a path connecting them. The* **distance** *between two vertices is the number of edges in the shortest path connecting them. For example, consider a and f. There are paths of length 2, 3, and 4 connecting them. Since 2 is the smallest, this is the distance between a and f. Note that if the graph models a physical situation, the distance on the graph is not the same as physical distance. Fun examples of distance include the classic "six degrees to Kevin Bacon", the mathematical version of this, "degrees to Erdös", and even experiments by Milgram on the subject (cf. [13]).*

> *Now suppose that instead the graph models conveyer belts between stations in a factory. Let every vertex represent a station that a product must visit in the factory before it is ready to leave the factory, and conveyer belts (or other automatized transportation for the product) between these stations be edges in the graphs. Remove j for this conversation. Are there any belts that are not needed? For instance, if belt bc is removed, then b is an isolated station and items could not reach this station on belts. However, observe that you could eliminate the belt cf and still get items to each station without carrying them. There are other belts we could choose to eliminate including: cd, df, de, and ef, though if we eliminated all of these our graph would have multiple components and there would be several stations that would require items be hand carried. But, we can eliminate more than one edge and still reach stations a-i with belts. For example, we may eliminate cf and de.*

It might help to draw the new graph or mark out these edges on the graph, and not erase them because they might be useful for conversation with the students later.

> *Notice that when we have eliminated all of the redundancy we are left with a graph that does not have any "circles". These internal circles are called* **cycles**. *More formally, a cycle is a path that begins and ends with the same vertex. We notice that the graph for the non-redundant conveyer belt system is not a cycle, contains no cycles, and is not simply a path since it has branches. This is an example of a* **tree**, *or a connected graph without loops such that any two vertices are connected by exactly one path. Notice that there are other trees inside of our full graph. For example, we can take the portion of the graph with vertices $\{a, b, c\}$ and edges $\{bc, ac\}$. This is a smaller tree (similar to our very first example). However, the tree we found was special among trees in this graph because it contains every node in the graph (except j, which we omitted). When a tree does this it is called a* **spanning tree**.

> *Let's revisit how we found our spanning tree. We eliminated redundant edges, or unnecessary belts, until we were done. This is one* **algorithm** *or procedure used to finding a spanning tree. In real life, though, all conveyer belts are unlikely to be equal. Maybe one belt is really old and needs lots of maintenance to keep it in working order. Maybe one belt takes a long time to deliver goods because it*

has to go up and over obstacles and moves slowly. Maybe one belt is awkward to move around and things frequently get caught in it making it dangerous. This illustrates that we may want to keep track of more information about each edge in our graph so that we can be more careful in our selection. In Figure 8.2, we add labels to the induced subgraph defined by $\{c, d, e, f\}$ since this contains all of the edges that are redundant. These labels are called **weights**, *and they keep track of some additional data related to the edge itself. Weights could represent any of the factors we mentioned before. Assume in this case that the weights are the total cost to run the belt for one day, including electricity and maintenance.*

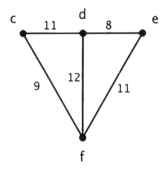

Figure 8.2. Edges with Weights

Now that we have this information, how might we choose which belts to eliminate? Previously, we eliminated cf and de, saving 17 dollars a day (assuming that the elimination does not increase the cost of the others substantively). Can we do better? Since our graph is small, we could use trial and error, but let's think about a systematic way to approach the problem. The edge df is the most expensive, so we eliminate it. Now we have two edges that could both be eliminated while still leaving a spanning tree, cd and ef. Thus our total savings, in either case, is 23, better than before. These trees with lowest total cost (more generally, lowest total weight) are called **minimal spanning trees**, *and the method that we used— eliminating expensive edges where there are cycles—is a reduction algorithm.*

Alternatively, instead of creating our tree by starting with the whole graph and eliminating edges, we could have started with just the vertices and selected edges from E of least cost to add such that each edge connects a new vertex and does not complete any cycle. In this example, we start with the addition of de since it has lowest weight. Now we add an edge that connects to d or e that has lowest weight, which is either cd or ef. Finally, with either choice, we add cf. Notice that in our example this new technique, called **Prim's Algorithm**, *gives us the same minimal trees as those found with the reduction algorithm. It is important to note that these algorithms do not always yield the same minimal spanning trees.*

While not necessary, this is a good place to introduce students to shortest path algorithms, such as Dijkstra's algorithm, or to introduce the concept of directed edges.

4.3 Putting the Two Together.

Introduction to Graph Theory Lab. To get comfortable working with graphs and networks at this stage, students complete an introductory lab that looks at interactions between classmates; Appendix A.1 contains a sample handout. This activity takes approximately one class period and is written to utilize class-collected data over a weekend. The instructor will need to have students collect and bring the data to class in advance. If the class is larger than 20, the instructor may wish to invent data to avoid dealing with a data set that is too large to be analyzed quickly by hand. This lab may reveal confusion about the algorithms. For example, some students may try to work vertex to vertex for the reduction algorithm rather than with the graph as a whole.

Note that to utilize class data students will need to record their number of interactions with each peer and bring this to class. In this case, an interaction means an opportunity to exchange information. This would include being in a class together, seeing each other on campus, texting, or even having a phone call. Data should be collected from the end of the class period to the start of the next. Essentially, students create a graph that models this network and use minimal spanning tree algorithms to determine a "best" method for disseminating information to the class.

Transportation of People Lab. After the first lab, students think about the movement of people in the context of human trafficking in a followup lab that takes five class periods. Options for shortening the lab are available in Section 5.2.

People that are trafficked can, unfortunately, be thought of as goods that are moved. However, while goods, they are still people and are therefore typically moved using existing networks for the transportation of people. In this lab, students begin by looking at airline maps, which tend to be freely available online from a variety of airlines. The report from the UNDOC indicates that in 2006 and 2007 most of the victims located in the United States came from Latin America, the Caribbean, and Asia. Assuming that these regions remain the primary origin of trafficked victims into the United States, students determine the most likely cities of arrival for trafficked victims. Students will notice that airlines have different hubs, minimizing the entry points for victims, making this portion of the problem relatively simple. Next, they examine highway maps. Assuming that after arriving in the United States, victims are moved via ground transportation since it is less rigidly controlled, they use their graph-theoretic knowledge to determine likely major cities of trafficked individuals. Students then seek to verify their findings by searching the news and data available from organizations such as *The Protection Project* [**10**] and *Polaris Project* [**9**]. Students are asked to reflect on their results and the implication of the findings for policy and other support for victims.

The lab is time-intensive with many tasks that need to come together before students can draw conclusions. This is intentional as one goal of this lab is to encourage effective team work by getting students to determine how they will coordinate the tasks given each person's strengths. Teams that work well together will finish without needing to complete any of the lab for homework according to the schedule provided in Table 8.1. However, teams that do not take the time to organize themselves well will struggle until they start to communicate. Students will be more likely to work well as a team if this is discussed in advance.

The graph directions for the airline portion leaves flexibility for vertices to represent individual countries, regions, or even cities. This can lead to fantastic discussions between students about the best way to choose this, and their decision will likely depend on the airlines that they choose since the airline maps tend to vary in this way. There are some domestic airlines that will have more flights between the regions of interest than others. If time is a factor, the instructor may wish to preselect the airlines that are most reasonable to use.

The lab is written such that the vertex set of the graph is the set of states, which allows one to consider multiple cities in each state. However, this can easily be rewritten to use highest degree cities (on the highway map itself), forcing specific cities to be used as vertices in place of states. Links to maps for this portion of the lab were placed on the course web site. It is vital that students read all of the directions before starting the highway portion of the lab. Instructors may wish to discuss this portion of the lab with students as a class after they have finished the highway portion.

A sample handout for the lab is provided in Appendix A.2. It is possible to split the lab and consider the ports of entry, or airline section, separate from the highway portion. Ways to shorten the labs are considered in the next section.

5 Additional Thoughts

5.1 Extensions and Connections. Notice that in this module we carefully use the mathematics to highlight and better understand the topic of human trafficking. The use of transportation networks to model the movement of goods is not novel. But it is this precise fact that highlights for students the lack of humanity attributed to those being trafficked. Moreover, returning at the end to look at the flaws in the real world data forcefully reminds us that we are indeed dealing with humans who do not speak out due to fear.

Another particularly nice feature of this module is the wide range of connections to be made to other areas of mathematics and social justice, as well as to the humanities. An issue commonly explored with networks is that of disaster planning, whose social justice connections became particularly clear after Hurricane Katrina. The Sahana Software Foundation is a useful starting tool for this extension [11]. Similarly, conflict networks and communication networks are obvious areas of possible connection. Another network of significance for students is the Internet itself, and there is much available in [1] and elsewhere on this topic, but this requires the introduction of directed edges. For those interested in adding more graph theory to the module, it would be easy to explore concepts such as sociograms, triadic closure, strong and weak ties, clustering, digraphs, flow, and other advanced topics [1,4,13].

Since human trafficking is considered the modern form of slavery, there is an organic parallel that students can investigate. For instance, the class could look at slavery as well as the post-slavery sharecropping culture in the South and ask:

- *In what ways is the current human trafficking situation like historical slavery in the United States?*
- *How does it differ?*
- *Similarly, compare and contrast the sharecropper culture with the human trafficking situation.*
- *Which situation is the most appropriate comparison?*

There are many natural opportunities for writing instruction in this module. Students could be asked to write about the text or further investigate current policies in a research paper. Alternatively, the final lab could be recast with less structure into an investigation culminating into an expository paper. A deeper paper would require students to intertwine information from traditional resources (like books and articles) with data and knowledge gained from the lab. Students could even be asked to make an infographic representing some of the data.

Another interesting approach would be to have students read excerpts from [12]. While this text is also titled *Sold*, it is a non-fiction account of one woman's experience being trafficked. Students can then compare and contrast the fictional account with the true story, as well as explore the differences in the styles of writing that appear in the two texts.

A different way to enhance reading and writing, as well as help students grow in Perry's scheme of intellectual and ethical development [8], is to implement the optional in-class debate after reading [7]. The instructor can make character cards for individual or groups of characters in the text, for example, for Lakshmi's parents or the tea boy. Students select a card when they enter the class and then are asked to argue that the character they selected is ultimately responsible for Lakshmi's enslavement. If there is time, the cards can be color-coded so that students with related characters can work together to construct their argument. Then the class can debate the question:

Who is ultimately responsible for Lakshmi's enslavement?

The process could be repeated to determine who is ultimately responsible for Lakshmi's escape. This debate generates lots of conversation about what we assume in a text about the characters and how we judge the different players. It also helps students develop the ability to take different perspectives.

The last portion of the Transportation of People lab, where students determine the validity of their results, is a natural place to introduce or reinforce information literacy. Students need to evaluate the validity of the sources they are using to support their results. Similarly, as one discusses ideas of validity and bias, there is space for discussing data collection and using these same ideas to better understand the significance of the reported statistics.

Finally, there is a less obvious connection that is within easy grasp—that of vocation and crafting one's path. The survivors choose what they do with the experience. Somaly Mam and those in the organization she founded choose to confront the difficult reality of their path and use that in a productive way that is clearly nothing short of a vocation. What lessons can students, who are starting to make their mark on the world, learn from survivors like these? How can they apply these lessons to their future?

5.2 Shortening the Module. The module discussed here is designed to take approximately one quarter of a semester, and could easily expand to fill a whole course, but instructors might not have this much time to devote to this topic in their course. The social justice topic can be primarily introduced using the DVD *Half the Sky* [3], which will reduce the introduction of the topic to simply one class period. It should be noted that this will provide a less layered understanding of this social justice topic, and care will need to be taken to ensure that students do not disconnect the topic from what they are doing mathematically.

The labs may be shortened in a variety of ways. One is to simply have students determine the major entry points using the airline maps. These are steps #1-5 on the "Transportation of People Lab" in Appendix A.2. One can also reduce the number of airlines that students need to look at, or assign each group only one airline to look at and conduct the full analysis as a class. The highway portion of the lab could be shortened by restricting the study to a region of the United States, though this will likely yield results different from the full lab. For instance, it might be interesting to look at what highways would likely be used to move people from a hub for one major airline, for example, from Atlanta, Georgia, to Florida. Students could then identify major cities along this short route between adjacent states. This would reveal that cities such as Macon, Georgia, are likely to be key in the movement of trafficked people. The same policy and law questions from the original lab could be asked about these cities. Each of these shortened labs could likely be completed in just over one class period.

Using a variety of these modifications, the unit can be completed in three or four class periods: introduce the topic of human trafficking, introduce graph theory, drop the introductory graph theory lab, and complete the shortened airline lab. Another class period could be used to complete a shortened highway lab. It is necessary to think carefully about the objectives instructors wish to achieve in class to determine the best implementation for their course.

Appendix A: Assignments and Handouts

A.1 Introduction to Graph Theory Lab.

Introductory Activity

In this activity you further explore the graph techniques we learned over the past week to explore a basic communication network. For this activity, you may work alone or in partners.

(1) On the board there is a chart to record how many times you interacted with each person in the class this weekend. Find the row with your name and put the number of interactions for each person in the appropriate column. Note: If Alexey has recorded 8 interactions with Sarah, then Sarah should have 8 with Alexey. If you find your data disagrees with someone else's, talk with them and resolve the issue. Explain: How does this table relate to an adjacency matrix?

(2) Utilize the table on the board to create a graph of class interactions over the past weekend. The set of people in the class are your vertices and the edges indicate interaction. If there is no interaction, then no edge should exist. If an edge exists, assign its weight to be 1 divided by the number of interactions. Decide: Does a higher or lower weight indicate more interactions?

(3) Suppose that I forgot to make an important announcement in class but email is not working. I want to get information to each of you quickly. Whom do you believe I should give information to first? Why?

(4) Are there any isolated vertices? What does this mean for the spread of information? For the remainder of this lab remove the isolated vertices.

(5) Let's assume that maximizing the number of interactions will minimize the amount of time it takes for everyone to receive the message. To do this, we want to find a minimal spanning tree of our graph. Use the elimination method to find the minimal spanning tree. What is the total weight of the tree?

(6) Now use Prim's method to find the minimal spanning tree. Is the tree the same as before? If not, state what has changed. What is the total weight of the tree?

(7) Looking at the two minimal trees, find the degree of each person. Is it reasonable to assume that if the person in each tree with the highest degree receives the information first that this will be the fastest way for everyone in the class to get the information? Why or why not? Compare your thoughts here to your answer in (3).

(8) Now suppose that instead of our class, this network is the communication network for an organization of human traffickers. What could you conclude using the graph? Who should a law officer investigate to learn the most about the organization the quickest? Explain your rationale.

A.2 Transportation of People Lab.

Transportation of People Lab

In this lab you apply the techniques that you have learned in this unit to determine the most likely hubs of human trafficking. As we know, data on human trafficking is sparse and unreliable; thus we use the theory of graphs and networks developed in class instead. For this activity, you may work in teams of 3-4.

(1) According to most recent UNDOC reports, most human trafficking victims to the United States originated from Latin America, the Caribbean and Asia. First we need to locate flight maps online. Locate the flight (or route) maps for at least: 2 United States airlines, 1 Mexican airline, 1 other Latin American but not Caribbean airline, 1 Caribbean airline, and 2 Asian airlines. Record your airline choices and their respective locations.

(2) Create a graph for flights entering the United States. The vertices inside the U.S. will be arrival cities. The vertices outside of the U.S. will be countries, cities, or regions (your choice). Each flight connects a vertex outside of the U.S. to one inside. Add undirected edges to represent these incoming flights.

(3) Record observations about your graph. Is there anything that surprises you about your graph? Is your graph connected? Are there high concentration areas?

(4) Find the degree of each vertex and determine the top four hubs for the transportation of people into the United States. What do you notice about the hubs? Are they in similar regions of the United States? Are any cities hubs for multiple countries? Are there any source countries not represented by your hubs? If so, eliminate any redundant hubs and add new ones to include these areas.

(5) Suppose that a future study indicates a rise in U.S. human trafficking victims from southern Africa since the UNDOC indicates that there are few specific anti-trafficking laws in this area. Would this change any of your hubs? Why or why not?

(6) **Read this entire item first, including all the examples below, and ask if you need clarification before proceeding.** Now that we have the hubs for point of entry, let's look at major cities across the U.S. that may be centers for human trafficking. We make a few simplifying assumptions. (1) There is a desire to traffic victims from the four hubs to each of the 48 contiguous states. (2) If people are

moved through a state, you would stop in the state to leave some people. (3) All stops will occur at major cities.

Download the U.S. highway map, and pick up a blank map of the U.S. In the new graph, each state is a single vertex. All cities within a state will be considered part of this vertex. Let edges represent major highways (or combinations of highways) that can be used to travel between states. There may be multiple edges connecting states, as illustrated below. All travel will be between cities large enough to be listed on the downloaded map (because of assumption 3). Assign weights to be the approximate mileage to travel between the states via the specified route. Utilize online map programs to find the mileage.

For example, the highway map shows that you can take Highway 5 to travel between Portland, OR and Olympia, WA. This means there is an edge between OR and WA with weight 114 miles. However, you can also travel between Portland, OR and Spokane, WA using Highways 84 and 90. This will add another edge to your graph with different mileage.

For states containing a hub, however, you must depart from this hub and thus it is the only city we will consider in that state. For example, if Tucson is an airport hub, all travel to and from Arizona must begin or end at Tucson.

Draw your graph. Notice that due to assumption (2), from any given state you only need to look at paths to adjacent states.

(7) Create a minimal spanning tree for your map using the elimination algorithm. Assuming that traffickers only use major highways when transporting people, the minimal spanning tree minimizes the number of miles to each city. What are the major highways of importance? What states end up being most significant? Do the cities with the highest degree remain?

(8) Use Prim's algorithm to create a minimal spanning tree. Compare this to your previous spanning tree.

(9) Summarizing the results of the previous two questions, identify the cities that are major hubs for each state. Note: This means thinking carefully about which route each edge represents. Is it Portland or Olympia that matters in Washington?

(10) You have now identified likely centers for human trafficking. Further, you have identified likely paths for moving individuals between these hubs. How valid are your results? Utilize websites like www.polarisproject.org, internet searches, and library databases to search for maps, articles, or sources on human trafficking in the cities you have identified. Does the limited available evidence support your findings? Does it support one of your minimal spanning trees more than another? Explain.

(11) Now that you have likely locations for the entrance of victims into the country, paths for trafficking, and state centers, what implications are there for policy, law, and other support services in these areas? Be specific.

(12) As a team, write a one to two page typed reflection that summarizes your results, their validity, the assumptions of this model, and the implication of your findings.

Bibliography

[1] David Easley and Jon Kleinberg, *Networks, crowds, and markets*, Cambridge University Press, Cambridge, 2010. Reasoning about a highly connected world. MR2677125

[2] N. Kristof and S. WuDunn, *Half the Sky: Turning Oppression into Opportunity for Women Worldwide*, Vintage Books, New York, 2009.

[3] M. Ferrari, M. Chang, J. Bennet, M. Chermayeff, N. Kristof, et al., *Half the Sky*, Docurafilms: New Video Group, New York, 2012.

[4] Daniel A. Marcus, *Graph theory*, MAA Textbooks, Mathematical Association of America, Washington, DC, 2008. A problem oriented approach. MR2503650

[5] Earlham College, *Principles and Practices*, `http://www.earlham.edu/about/mission-beliefs/principles-practices/`, accessed on May 10, 2016.

[6] Online Atlas, *United States Interstate Highway Map*, `http://www.onlineatlas.us/interstate-highways.htm`, accessed on May 10, 2016.

[7] P. McCormick, *Sold*, Hyperion Paperbacks, New York, 2006.

[8] G. Perry, *Forms of Intellectual and Ethical Development in the College Years*, Holt, Rinehart and Winston, Inc., New York, NY, 1968.

[9] The Polaris Project, `http://www.polarisproject.org`, accessed on May 10, 2016.

[10] The Protection Project, `http://www.protectionproject.org`, accessed on May 10, 2016.

[11] The Sahana Software Foundation, `http://sahanafoundation.org`, accessed on May 10, 2016.

[12] Z. Muhsen, *Sold*, Little, Brown and Company, London, 1991.

[13] M. van Steen, *Graph Theory and Complex Networks: An Introduction*, M. van Steen, Amsterdam, 2010.

[14] D. Walker, *God In A Brothel: an undercover joinery into sex trafficking and rescue*, InterVarsity Press, Downers Grove, 2011.

[15] United Nations Office of Drugs and Crime (UNODC), *Global Report on Trafficking in Persons*, February 2009. Available at `http://www.unodc.org/unodc/en/human-trafficking/global-report-on-trafficking-in-persons.html`, accessed on May 10, 2016.

9

Evaluating Fairness in Electoral Districting

Geoffrey Buhl and
Sean Q Kelly

Abstract. In this module, students are introduced to aspects of "fair" electoral districting plans and develop mathematical tools to evaluate how well districting plans conform to normative conceptions of fairness. Students are asked to formulate personal notions of fairness in districting, to district a city, and to evaluate their districting plan for fairness using mathematical tools. In a hands-on activity appropriate for any college student, students create voting districts for a hypothetical city with a heterogeneous population and varied political affiliations. They then use mathematical tools to evaluate the fairness of the district lines that they draw. Through this activity, students experience the real challenges of districting and see how different notions of "fairness" can tug districting lines in opposite directions. The mathematical content includes two measures of compactness for planar shapes and a measure of voting power. **Keywords:** Shape factor, plane geometry, fair division, electoral districting, group work, in-class discussion.

This module was developed for a team-taught interdisciplinary general education course and contains scaffolded activities for two to four class periods that can work in any general education, liberal arts, or quantitative literacy mathematics course. The content has a connection to discrete mathematics and geometry and thus can be used in those courses, too. The lesson of this module is that mathematics is a tool that can help us evaluate normative conceptions of fairness. The setting for this exploration involves electoral districting plans for a hypothetical city, Squareville, with simplified geometry that enables hands-on group work.

1 Mathematical Content

The activities and concepts in this module are appropriate for any college student. The mathematical concepts at the heart of this module are: two area-to-perimeter measures of compactness based on the isoperimetric inequality and a measure of voting power for voters in elections with a non-uniform distribution of votes.

1.1 Measures of Compactness.
Compactness in the redistricting setting attempts to describe the idea that shapes of electoral districts should not be too stretched out nor should the boundaries look too undulating or jagged. For the first measure, Polsby-Popper [1] gives a quantification of the jaggedness of a planar shape's boundaries. For the second measure, Reock [2] quantifies the dispersion or oblongness of a shape.

Both compactness measures are based on the isoperimetric inequality [3]. The isoperimetric inequality states

$$L^2 \geq 4\pi A,$$

where A is the area enclosed by a planar curve of length L. The only shape for which equality holds is a circle. Circles do not suffer from dispersion; the area is uniformly distributed around the center. Circles have a boundary with constant curvature and do not exhibit contorted boundaries. For this reason, for both measures of compactness, district shapes are compared to a reference shape that is a circle.

To make the activities hands-on, district shapes are built out of unit squares lying on a square lattice. The compactness measures, designed for planar shapes, are translated into this unit square shape, or polyomino, setting.

For polyominoes, a simple argument shows that no concave district can maximize the area-to-perimeter-squared ratio, using a fill-in-the-indented corner construction. If a polyomino shape is concave, there will be an opportunity to add a unit square to shape that preserves the total perimeter of shape while adding one to the total area. This shows that the polyominoes that maximize the area-to-perimeter-squared ratio must be rectangular. Among rectangles, squares maximize the area-to-perimeter-squared ratio. The isoperimetric inequality for square lattice shapes is

$$P^2 \geq 16A,$$

where P is the perimeter [4]. For the calculations of compactness for polyominoes, our reference shape is a square.

Polsby-Popper Measure. The Polsby-Popper measure of compactness compares the area of a connected planar shape to the area of a circle with the same perimeter as the circle. Thus it is a measure of the jaggedness of the boundary of the shape. For planar shapes, if P is the perimeter of a shape S, then the Polsby-Popper measure is

$$\frac{\text{area}(S)}{4\pi P^2}.$$

Close to 1, the boundaries are close to constant curvature, and as the measure decreases, the more contorted the boundaries become.

In the square lattice geometry we use, the perimeter of district shapes is easy to calculate. In order to keep the same zero to one scale, the formula for the modified Polsby-Popper measure becomes

$$\frac{16 \text{ area}(S)}{P^2},$$

mirroring the isoperimetric inequality for square lattice shapes.

Reock Measure. The Reock measure of shape compactness compares the area of the shape to the smallest circle that the shape can fit in, i.e., the circumscribing circle, and measures the dispersion of a shape. In particular, given a shape S with circumscribing circle C, the Reock measure of compactness is the ratio of the areas

$$\frac{\text{area}(S)}{\text{area}(C)}.$$

This ratio measures dispersion or how much a shape is spread out from its center. If the Reock measure is close to 1, then there is little dispersion, while the smaller the Reock measure, the more dispersion the shape exhibits.

In the simplified setting of square lattice shapes, to find the modified Reock measure one only needs to find the circumscribing square or the smallest square in which the shape fits. To find the circumscribing square, determine the maximum of the height or width of the shape. The circumscribing square is the square with side length equal to this maximum. So the modified Reock measure for dispersion of a square lattice shape S is

$$\frac{\text{area}(S)}{d^2},$$

where d is the maximum of the height and width of S.

1.2 Banzhaf Power Index. The Banzhaf power index (BPI) is used when there are entities that control different numbers of votes in an election with different numbers of votes [6]. For example, these can be states in the electoral college or shareholders in a corporate vote. The propositions voted on must be simple propositions with two options. The BPI for a given voter is the ratio of the number of times that voter is a swing voter in a winning coalition to the total number of swing votes in winning coalitions. For this module, a *winning coalition* is a collection of votes that have 50% plus one or more votes among the members of the coalition. A *swing voter* in a winning coalition is a voter for whom if their vote changed, the coalition would no longer win.

For example, consider a district composed of citizens from four different communities of interest K, L, M, N. In this particular district, community K has 15 votes, L has

7, M has 4, and N has 5. We represent this problem as follows:

Coalition	Votes	Swing Votes
KL	22	K, L
KM	19	K, M
KN	20	K, N
LMN	16	L, M, N
KLM	26	K
KLN	27	K
KMN	24	K
KLMN	31	

In this example, a winning coalition must have 16 votes. How much power does each community have in this district?

There are 12 swing votes across all the winning coalitions. In terms of BPI, K has 6/12, L has 2/12, M has 2/12, and N has 2/12. If the threshold of the voting power that preserves a community of interest is a BPI of 50% or more, then this district does so for community K.

2 Context / Background

The activities in this module challenge students to examine their normative conceptions of *fairness* by examining different concepts in "fair" districting plans. Mathematics is unable to provide an absolute conception of fairness, but it plays a key role in evaluating to what extent particular conceptions of fairness may be achieved.

2.1 Political Context. Electoral districting is the process of creating geographic subdivisions within an electoral unit (e.g., a country, state, or city) for conducting elections for representation to a legislative body like the United States Congress, a state legislature, or a city council. Representatives to the legislature are chosen to represent the interests of the voters that elected them to office, and their reelection is presumably contingent upon successful representation of those interests as judged by the voters in subsequent elections.

States establish the process for drawing statewide electoral districts. In a majority of states the process is retained by the legislature; state legislatures — in some states there is a role for the governor — draw state legislative boundaries. In other states a commission draws the districts. Similarly, drawing electoral districts in municipalities may be done by a city council or a commission.

Five values figure into drawing district lines: equal representation, contiguity, compactness, electoral competition, and maintaining communities of interest.

Since the 1960s one value is inviolable: equal representation. In 1964 the United States Supreme Court ruled in the case *Reynolds v. Sims* that state legislative districts must contain equal numbers of people, establishing the principle of "one person, one vote". Prior to the *Reynolds* decision, for instance, the California State Senate was composed of forty members representing its 58 counties. As a result the one Senator from Los Angeles County represented 6 million voters while a Senator from a set of lightly populated rural counties represented fewer than 15,000 voters. A vote in Los Angeles County was "worth less" than a vote in the smaller rural district; one vote in an electorate of 15,000 had a better chance of determining the outcome of an election than one vote out of 6 million in another. Currently within a state every congressional district has the same population, so far as can be determined by census data.

Beyond equal representation, states typically impose other standards for drawing electoral districts. Forty-nine states require that districts be contiguous and twenty-three states require that districts are compact [5]. In this context, contiguous effectively means path connected, with exception for geographical features like islands. Definitions of compactness vary from state to state. For example, Idaho state laws state, "To the maximum extent possible, the plan should avoid drawing districts that are oddly shaped". Oblongness and the degree to which boundaries are contorted or jagged are frequently used a markers of "oddly shaped" or gerrymandered districts.

District lines can be drawn to increase the likelihood of a particular electoral outcome by *packing* voters with similar interests into a district to diminish their voting power outside the district, or *cracking* similar voters into different districts to dilute their influence in any district election. The purpose of both tactics may be to ensure the election of a specific individual politician or of a representative from a particular party or a particular group. This arrangement is referred to as a "gerrymander".

Gerrymandered districts are intended to reduce competition in a single district [7, 8]. Political reformers frequently contend that non-competitive elections cause politicians to be less responsive to their constituents, or that voters lose interest in the electoral process when outcomes seem predestined and their vote does not matter. Because non-competitive elections can have these effects, reformers commonly argue that the districting process should be performed by independent commissions rather than politicians who may have an interest in reducing competition.

Maintaining communities of interest is another value associated with drawing district lines. Contiguous and compact districts are often considered sufficient to maintain the integrity of interest representation; people with common interests often live in similar geographic areas. Shared interests, however, do not always conform to geography. In the 1980s, for example, the Reagan Justice Department began to push for the creation of majority-minority districts in states with a history of racial discrimination. Because these districts would be composed of a majority of minority voters (African-Americans and Latinos), a minority representative could likely be elected, thereby incorporating the interests of minority voters into the halls of government.

Any particular districting plan may require trade-offs between the values in this discussion. Compact and contiguous districts do not ensure competition or automatically maintain communities of interest. Promoting competition and interests may mean creating districts that are not compact and whose boundaries appear jagged in an attempt to maintain contiguity. Mathematics can shed light on each of these values — which ones are maximized, which are minimized — even while mathematics cannot answer the normative question of which values are more important to maximize.

2.2 Institutional Background. This module was developed for students in a general education course at an open-access state university. Students worked through this module in two settings: over the course of two weeks in a team-taught interdisciplinary general education course and a one-day three-hour workshop associated with an Environmental Science mapping course. The content of this module evolved while teaching an interdisciplinary course at the intersection of mathematics and political science.

2.3 Social Context. In this module, students develop basic knowledge of electoral districting in the United States. This is a significant step in understanding how this country is governed. Thus the module contributes to informed civic engagement of future generations.

The motivating questions of the module are:

- *What are some normative conceptions of "fairness" in different electoral districting?*

- *How can we use mathematics to evaluate how well a redistricting plan conforms to these conceptions of "fairness"?*

Districting provides a rich area for students to discuss their own notions of fairness. In a simplified geometric setting of this module, students engage in districting a city and then use mathematics to evaluate whether their districting plan conforms to their notions of fairness. In order to make the activity and the mathematics more accessible, the geometry of the city is simplified to an array of unit squares, each with three attributes: population, political affiliation, and community of interest. The simplified geometry allows students to use basic formulas and simple arithmetic to evaluate concepts like compactness, competitiveness, and preservation of communities of interest. Because the city's population is heterogeneous, students can explore the interplay between notions of fairness in districting plans.

3 Instructor Preparation

3.1 Before the Class Meeting. There are two important decisions to make before this module is brought into the classroom. First, the scope of the module is variable. Second, this is a hands-on districting activity, and so instructors need to decide how they will generate and distribute copies of the model of the city to be districted.

The first decision an instructor must make is how complicated a problem she wants the students to work on. In order of increasing complexity, the following problems fall within the scope of this module:

(1) Can we district this city into districts with equal populations, so that each district is reasonably compact?

(2) Can we district this city into districts with equal populations, so that each district is reasonably compact *and* each district is electorally competitive?

(3) Can we district this city into districts with equal populations, so that each district is reasonably compact, each district is electorally competitive, *and* communities of interest are preserved?

Depending on how much class time is dedicated, the instructor can determine an appropriate level of complexity for the hypothetical city.

The second critical element is preparing versions of a city for student groups to district. An important aspect of the students' experience of this module is to have some hands-on object that they can draw districts on. This could be a printed handout of the city to be districted or a spreadsheet with the city information in the spreadsheet cells. If students have access to a computer, a spreadsheet that allows color coding of information is a nice option.

For the districting activity, the most basic setting is a square city with blocks of varying population. The next level of complexity includes city blocks with population numbers and overall political party affiliation of the block (we have used Democrat, Republican, or Independent). The highest level of complexity is a city block with population size, party affiliation, and interest group membership. If students are using printed handouts, printed in black and white, each city block would have information about size, party, and group.

Appendix A.2 presents an example version of a hypothetical city. In the city, each city block has one of three political affiliations Republican, Democratic, and Independent. The number in each cell represents the population of the associated city block. The I, R, or D indicates the political affiliation of the block, Independent, Republican, or Democratic. The second letter on the bottom (L, M, N, or K) of each cell represents the community of interest of that city block. In the Mathematica code that we use to generate cities,[1] there is a 20% chance that a block will be Independent, and a 40% chance each that the block will be Democratic or Republican.

Information presented in black and white can be challenging for students to work with; coloring the population number to indicate political affiliation and the cell to reflect community of interest might be helpful. We typically use color printouts or spreadsheets with information color-coded for the districting activity. Sample cities of various difficulty levels for the districting activity are available at: `http://faculty.csuci.edu/Geoffrey.Buhl/districting`.

3.2 During Class. This module gives students an opportunity to come to a group decision on which aspects of fair districting are important, and then district a city guided by their collective conception of fairness. Exploring normative conceptions of fairness can feel out of place in a mathematics class. The value of this module is to learn mathematics that can help to evaluate fairness and to understand the limits of mathematics in defining fairness. Discussion of the normative aspects of districting provides context for the value of mathematics but also its limits. Providing class time for both the conceptions of fairness discussion and the mathematics is critical for students to fully absorb this module.

In our experience, the most challenging part for students is the normative element: selecting and justifying a notion of fairness in electoral districting. It is challenging to engage students in a discussion on important aspects in fair districting. For example, we might start with questions such as

- *For congressional districts, each district must have the same number of people residing in the district. Is this important to your concept of fairness in districting or is it unimportant? Why?*

This is difficult for students who may prefer to be told what is fair rather than decide or state what they believe is fair.

In our minds, the questions *What is fair?* and *Can we use mathematics to evaluate fairness?* are of equal importance in this module. Of course we could save time by simply stating, "This is what the federal government or states say is fair". But half of the value, we believe, of this module is the discussion that follows the question

- *Which of these conceptions of fairness are important to you, and why?*

---

[1] Sample code will be available online soon.

3.3 Post Assignment: Grading Issues. The culminating assignment prompt can be found in Appendix A.1. There are two fundamental questions that students should address in their work in this module. Students should answer both of the key questions posed by this module.

- *What is your concept of fair districting and why?*

- *How well does your districting plan fit with your concept of fair districting?*

In our team-taught course, this assignment reflects 25% of a student's course grade. We have graded the students' response to the two key questions equally, each question representing half the total grade on the assignment. In our experience, students generally do a good job using mathematics to evaluate their districting plan but have difficulty connecting the results of their calculations to their version of fairness.

4 The Module

In Subsection 3.1 we listed three levels of complexity for the module. Depending on how the instructor decides on this issue, the module will take as little time as one class period or as much as four. In particular, if the instructor chooses to focus on the first question, addressing only compactness, a single class period will suffice for the whole project. If the instructor chooses to pursue the second question, addressing compactness and competitiveness, the project may take two-to-three class periods. The most in-depth version takes three-to-four class periods.

Here is the motivating narrative for the module:

> The city of Squareville with diverse political affiliations and peoples, is moving from at-large elections for city council seats to district elections for city council seats. The goal of this change is to have the city council members better represent the interests of the citizens of Squareville.

In this context, students assume the role of a citizens' redistricting committee. As commissioners, students first decide on the qualities of a fair districting plan. Second, they create a districting plan for Squareville according to the their chosen principles of fairness.

There are five aspects of fairness in districting Squareville that students can explore: equal representation, contiguousness, compactness, competitiveness, and preservation of communities of interest.

When we use this module, the first two aspects, contiguous districts and equal populations in each district, are given as requirements. The latter three are explored in detail in the module and can conflict with one another. This tension between the latter three aspects of fair districting parallels the challenge of districting in the real world.

In teaching this module, a reasonable definition of a contiguous district is a district in which each square of the district shares at least one side with another square in the district. Because each city block cannot be subdivided, it is not usually possible to have all districts equal the average district size. For this module, a reasonable goal is for each district to be within 10% of the total population divided by the number of districts.

It is helpful to provide some readings for the students before the module begins. We suggest using [7, pages 50-89] and [8, pages 107–137].

4.1 Districting for Compactness. We suggest starting with a discussion preceded by an interactive lecture. Here the instructor may start out by explaining some of the following:

> *As many as 23 states require that their legislative districts be reasonably compact. However, there are nearly as many conceptions of compactness as there are states that require it. There are broad notions of what it means for a political district to be not compact. For example, if a district's boundaries are too "squiggly" or "jagged", then the district may not be compact. If a district is too spread out from its center or too oblong, then the district may not be compact. The first notion, contorted boundaries, and the second, dispersion, can be represented mathematically. Potentially complicating this, the Supreme Court, in a 2006 ruling on the Voting Rights Act, has indicated that cultural cohesion of a district may be considered a part of district compactness.*

Here are some motivating questions about district compactness that might be appropriate:

- *What is gerrymandering?*

- *Can a simple visual test identify gerrymandered districts?*

- *How can contorted boundaries be misused in districting?*

- *How can dispersed districts be misused in districting?*

There are websites that catalog gerrymandered districts, or at least districts that suffer from contorted boundaries, dispersion, or both [**9**]. Showing examples of such malformed districts helps inform potential abuses of non-compact districts.

Next the instructor may take some time to lecture, introducing the isoperimetric inequality, Reock measure, and Polsby-Popper measure. Students will also benefit from seeing how measures are translated into the polyomino setting.

At this point, students should be asked to work in groups trying to district a city with given population information so that individual districts are reasonably compact. Afterwards, the class could be asked to make a quantitative threshold for "reasonably compact".

4.2 Districting for Competitive Districts in Elections. Here the instructor may start with a brief interactive lecture, explaining some of the following:

> *Some states, for example, California, require that districts not "favor or discriminate against an incumbent, candidate, or political party". Requiring that a district not favor a party means that the election in each district should be competitive among parties. For competitive elections, the goal is to limit the power of partisan coalitions. In a competitive district both parties should have a nearly equal chance of winning an election. The other alternative, partisan districting, creates safe districts for certain politicians and parties, that is, districts where a certain entity is extremely likely to win election.*

> *A good example of partisan districting is Pennsylvania's congressional districting after the 2010 census. In 2012, Democratic congressional candidates received 50.3% of the votes cast but captured only 5 of Pennsylvania's 18 seats in the U.S.*

House of Representatives. By packing Democratic votes into a small number of districts following the 2010 Census, Pennsylvania's Republican-controlled legislature ensured that 49% of the vote would yield Republicans 72% of Pennsylvania's representation in the House.

Some good discussion questions here are:

- *Should districts be competitive or should there be safe seats for certain parties?*

- *On average, how closely should the ratio of elected officials, Democrat to Republican, match the ratio of registered voters for those parties?*

This module uses the Banzhaf power index to determine if districts are competitive or not. This is a good time to introduce students to this index and show how it can be used to analyze the competitiveness of electoral districts. For the purposes of this module, we assume Democrats, Republicans, or Independents vote in blocks, and Independents vote for either the Democratic or Republican candidate. A competitive district is a district where each political affiliation holds one third of the voting power. A partisan district is where one party holds all the voting power, that is, a BPI measure of 100%.

Then comes the next opportunity for group work. Students are asked to district a city with both population and political affiliation information given. Students can be prompted to district for competitive or partisan elections. In our experience, students enjoy districting for partisan elections and trying to maximize the number of safe districts for one party or another.

4.3 Districting to Preserve Communities of Interest. Here the instructor may start with a brief interactive lecture, explaining some of the following:

Another value that appears in political districting is: districts should attempt to keep intact communities of interest. The definition of communities of interest, which vary from state to state, generally attempt to qualify the characteristics of communities that should be considered when districting. For example, Colorado law describes factors in identifying communities of interest as "ethnic, cultural, economic, trade area, geographic, and demographic". Here we will simplify the complexity of human diversity by assuming that city blocks are members of only one community of interest.

Some good discussion questions here are:

- *Why is it important that the makeup of the city council be representative of the people in the city?*

- *Why is it valuable for communities of interest to have districts in which they have a large amount of voting power?*

- *At what size as a percentage of total voters should a particular community be considered large enough to warrant a district where it controls a sizable amount of voting power?*

- *What is the voting power threshold at which one can say that a district preserves a community of interest?*

- *If a community of interest makes up 40% of the city's population, should it have 40% of the city council districts allotted to it?*

- *Are political parties communities of interest? Why or why not?*

In evaluating how well a districting plan preserves communities of interest, the above can be followed up with asking students if the city council is representative of the communities that make up the city. This is another opportunity to create a mathematical measure for the political science idea of representation. One such measure could be described as follows:

> *If the proportion of districts drawn to preserve communities roughly matches the proportion of the populations among the communities, then the overall districting plan preserves communities of interest.*

Next students can be divided into groups and provided a city to district into a given number of city council districts. Once in a group, students construct their normative conception of a "fair" districting plan. They decide on a metric for compactness, districting for competitive or partisan districts, and a threshold of voting power which ensures that a district preserves a community of interest. Students then divide Squareville into districts that are contiguous, have equal populations among districts, and conform to their conception of "fairness" as closely as possible. After drawing the lines, students evaluate districts for their compactness, competitiveness, and preservation of communities of interest. They then use their findings to evaluate the overall fairness of their districting plan. This evaluation should describe the choices and tradeoffs the group made in creating the districting plan.

5 Additional Thoughts

The mathematical tools developed in this module are used to provide evidence for judgments about how well a particular districting plan for Squareville conforms to stated ideals of fair districting. The value in this module is twofold: the application of mathematical concepts to help answer questions about fairness and the formulation of notions of fairness in districting.

5.1 Further Investigations. The connections between mathematics and political science in redistricting are deep and rich, and by design, this module simplifies much of the mathematical complexity for the sake of accessibility. Below we list a few ideas and references that introduce some of this richness for more mathematically inclined students.

- The Banzhaf power index has a connection to generating functions [10]. Exploring algorithms for calculating the index efficiently is a question that a student of mathematics or computer science may explore.

- As defined in this module, the Polsby-Popper measure has some shortcomings. For example a district made up of three (or any other non-square integer number of) unit squares can never have an isoperimetric quotient equal to 1. Modifying the Polsby-Popper measure to account for district areas that do not have integer square areas is related to finding polyominoes of minimum perimeter for the given area. Students can explore what these minimal perimeter polyominoes are and use those shapes to create a table of best isoperimetric quotients for a given district with area k [11].

- Another concept used in districting is that of natural boundaries [12]. Natural boundaries may refer to political boundaries like city or county lines, they may be major streets, or they may be geographic boundaries like rivers. By adding some natural boundaries and weighting how the perimeter of the natural boundary counts, one can add a fourth level of complexity to this module.

- There are many other measures of compactness that are based on mathematics beyond the isoperimetric inequality and can be found in mathematical literature, e.g., [13], and political science literature, e.g., [14]. Cataloging and categorizing the various measures could be the basis for a project that exposes students to searching for and reading mathematical texts.

Appendix A: Assignments and Handouts

This appendix contains two examples of assignments we used with this module. The first was used for the culminating assignment in a team-taught upper-division interdisciplinary class, with multiple weeks of the course dedicated to this module. The second was used in a three-hour seminar in an Environmental Science mapping course. Both assignments used a similar 10×10 grid city; see the second appendix.

A.1 Culminating Assignment Prompt. We used this module in the latter portion of an interdisciplinary general education course, which fed into the students' final assignment for the course. Below is the prompt that we gave students along with their own version of a 10×10 city to district. This paper represented a quarter of their course grade.

> *The first part of this assignment should be a description of your personal views on what districting priorities result in "fair" districts. This will include your definition of what a fair districting scheme is or looks like. You then need to take up the question of "fairness" squarely. Do you think your plan is fair? Explain. What would opponents of your plan argue about its lack of fairness? Do you think there is a conception of fairness that is possible in the districting process? Why do you think that? Draw on the Brunell [7] and the Bullock [8] readings to support your discussion of fairness.*

> *The second part of this paper will entail the districting of your city into seven districts and a numerical analysis of your districts for partisanship, voting power, compactness, and the proportional representation of communities of interest. Using the numerical data for your districting scheme, you should analyze how successfully or unsuccessfully your districting scheme meets your own criterion for fairness. You should qualify how well your seven districts represent the population distribution of your subpopulations.*

A.2 Example Handout. The purpose of this assignment is for you to district a city according to your personal priorities of competitiveness, partisanship, communities of interest, and compactness. In your write-up you will analyze your districting scheme and justify the choices you made. You will also describe your personal priorities for districting and explain how and why they are important to you.

You will district the attached 10×10 city into 7 districts that must satisfy the following basic districting rules:

(1) Each district must have approximately the same number of people. A 10% margin for error is allowable.

(2) Each district must be contiguous. Squares in a district must share a side with another square in the same district. You cannot have isolated square or squares that only touch at a corner be in the same district.

(3) Use a measure of compactness on your districts.

(4) Provide a voting power analysis on each district. This should include an analysis of the voting powers of communities of interest as well as political affiliations.

1 D K	4 R M	2 I N	3 D M	4 D M	5 D M	3 I K	1 D K	4 D K	4 R M
1 D K	3 D M	2 D M	3 D M	3 R L	5 R K	2 D K	2 D K	4 D K	5 R N
1 R K	4 D K	1 D M	4 I K	4 R K	2 I M	5 I M	4 D L	3 R K	1 D N
2 D M	1 R K	3 D M	1 D M	3 D K	3 R K	2 R K	4 D M	2 R N	5 R L
1 R K	1 D K	5 R M	3 R N	5 I K	2 R K	2 R K	1 D N	2 D M	1 R N
2 I M	1 R N	1 D M	2 D M	3 D L	4 I N	4 D N	3 D N	1 D N	1 D N
1 R M	1 R M	2 R L	2 D K	2 D M	4 D L	5 I N	1 R N	1 I N	1 D N
1 R L	2 D M	2 D K	4 I L	1 D L	2 R L	5 R N	2 R K	5 D N	1 D N
1 R L	1 I L	1 R N	3 I N	5 D L	5 R N	5 D N	3 I N	4 R N	2 D K
2 I L	4 D L	2 R L	1 I L	5 R L	3 I L	4 D L	5 I N	1 R L	1 D N

The descriptive statistics for this city are:

Republican	91	K	67
Independent	55	L	58
Democrat	118	M	64
		N	75
Population	264		264

Bibliography

[1] D.D. Polsby, R.D. Popper, "The Third Criterion: Compactness as a Procedural Safeguard Against Partisan Gerrymandering," *Yale Law and Policy Review* 9 (1991): 301–353.

[2] Reock, Ernest C., Jr., "Measuring Compactness as a Requirement of Legislative Apportionment," *Midwest Journal of Political Science* 5 (1961): 70–74.

[3] Robert Osserman, *The isoperimetric inequality*, Bull. Amer. Math. Soc. **84** (1978), no. 6, 1182–1238, DOI 10.1090/S0002-9904-1978-14553-4. MR0500557

[4] Da Lun Wang and Ping Wang, *Extremal configurations on a discrete torus and a generalization of the generalized Macaulay theorem*, SIAM. J. Appl. Math. **33** (1977), no. 1, 55–59, DOI 10.1137/0133006. MR0438237

[5] Levitt, Justin, "Where the lines are drawn - congressional districts," *All about Redistricting: Professor Justin Levitt's guide to drawing the electoral lines.* `http://redistricting.lls.edu/where-tablefed.php`, accessed on June 1, 2016.

[6] Banzhaf, John, "Weighted voting doesn't work: A mathematical analysis," *Rutgers Law Review* 19 (1965) no. 2: 317–343.

[7] Brunell, Thomas, *Redistricting and representation: Why competitive elections are bad for America*, Routledge, New York, 2008.

[8] Bullock III, Charles S., *Redistricting: The most political activity in America*, Rowman & Littlefield Publishers, Lanham, MD, 2010.

[9] Ingraham, Christopher, "America's most gerrymandered congressional districts," *Washington Post*, May 15, 2014. `http://www.washingtonpost.com/news/wonkblog/wp/2014/05/15/americas-most-gerrymandered-congressional-districts/`, accessed on June 1, 2016.

[10] J. M. Bilbao, J. R. Fernández, A. Jiménez Losada, and J. J. López, *Generating functions for computing power indices efficiently*, Top **8** (2000), no. 2, 191–213, DOI 10.1007/BF02628555. MR1832638

[11] Yaniv Altshuler, Vladimir Yanovsky, Daniel Vainsencher, Israel A. Wagner, and Alfred M. Bruckstein, *On minimal perimeter polyminoes*, Discrete geometry for computer imagery, Lecture Notes in Comput. Sci., vol. 4245, Springer, Berlin, 2006, pp. 17–28, DOI 10.1007/11907350_2. MR2305651

[12] Birge, John R., "Redistricting to maximize the preservation of political boundaries," *Social Science Research* 12 (1983) no. 3: 205–214.

[13] Jonathan K. Hodge, Emily Marshall, and Geoff Patterson, *Gerrymandering and convexity*, College Math. J. **41** (2010), no. 4, 312–324, DOI 10.4169/074683410X510317. MR2682920

[14] Chambers, Christopher; Miller, Alan, "A Measure of Bizarreness," *Quarterly Journal of Political Science* 5 (2010): 27–44.

10

Modeling the 2010 Gulf of Mexico Oil Spill

Steve Cohen and
Melanie Pivarski

Abstract. In April 2010 the Deepwater Horizon oil rig exploded, beginning an oil spill that would last until July of the same year, with a final seal of the well occurring in September 2010 [5]. How can we estimate the amount of oil that was spilled? We explore areas, volumes, sums, piece-wise functions, and differential equations in a series of group projects to model this problem. Data come from the National Oceanic and Atmospheric Administration maps of affected areas [9], estimates of recovered oil [1], and estimates of flow from the well itself [5]. Students discuss what information is needed to build a reasonable model, identify sources of error, and discuss why systematic processes are useful. This project has been used over the course of a one-semester integral calculus course; however, individual pieces would be appropriate for differential calculus, mathematical modeling, differential equations, or general education mathematics courses. **Keywords:** Pollution, environmental modeling, areas and integration, differential equations, group work, communication.

We describe a six-part module that can be used in most integral calculus classes when spread over the entire semester. Alternately, the parts in this module can be used individually or in a shortened form. The final part is ideal for faculty members who would like to incorporate a communication component into their calculus course. Although the amount of oil spilled in the Gulf of Mexico is a well-defined quantity, it is impossible to measure it directly. The module design helps students understand how integrals and simple differential equations from a calculus course help one generate a model of a real situation where it is difficult to find a good dataset.

1 Mathematical Content

This multi-part project is designed for an integral calculus course, but it is also ideal for a mathematical modeling or differential equations course. It consists of five mathematical parts and an optional library part the first of which could be used in a differential calculus class to motivate the upcoming idea of Riemann sums or in a general education math course. Handouts for all parts are included in the appendix.

Parts 1 and 2 require very little background. Students estimate areas by splitting a region into smaller, easily measurable pieces. Students solve differential equations in Part 3, so knowledge of derivatives and some ability to integrate by hand or using technology is necessary. Students do exercises translating various scenarios into differential equations and solving them. Techniques used include: simple integration (3abh), separable differential equations (3cd, 4ab), linear differential equations (3gi,4ab), integration by partial fractions (4b), and expressing solutions in terms of unknown functions. For example, in exercise (1e) students determine that $\frac{dy}{dt} = A - f(t)$, and in (4c) they find that the solution is $y = At - \int f(t)dt$. Part 4 entails fitting a curve to the data, and students will need to use technology, such as Maple, Mathematica, Matlab, or (for simpler fits) Excel. In the final part, students write or present a poster of what they discovered using mathematical communication skills.

2 Context / Background

2.1 The Deepwater Horizon Oil Spill. On April 20, 2010, an explosion erupted aboard the Deepwater Horizon platform. Oil began to leak on April 22 when the riser fell to the seabed; the measures in place to contain the oil failed. For eighty-five days, oil leaked into the Gulf of Mexico until the well was contained on July 15, 2010 [5].

During this time, many different groups worked to model what was happening, including the rate with which oil flowed out of the well, how the oil was spreading and breaking down, and the surface area affected. Estimates of the flow rate varied wildly, from one thousand to 70 thousand barrels per day (BPD) [1, 12]. After further study, the estimate was found to be between 48,000 and 66,000 BPD with the amount varying over time [5]. A good estimate for the flow rate was needed to set the engineering specifications for containment, as well as to determine the total amount of oil leaked into the ocean. This total was then used to estimate the amount of dispersant needed to minimize damage. Estimates of surface areas of the spilled oil were needed for ships and wildlife [5].

Many methods were tried to contain the oil: use of a blowout preventer (which did not work due to the depth at which the spill occurred), multiple tries with riser insertion tubes to help capture oil from the well, and attempts at capping the well. On June 3, 2010, the riser insertion tube was cut, which led to a small increase in the flow rate. Relief wells were drilled, a new blowout preventer was installed, and the well was eventually capped. The bulk of the flow was contained by July 15, 2010 [1]. By September 16, 2010, the well itself was considered to be dead.

2.2 Finding Data. Given the nature of the spill, it was difficult to obtain accurate data on how much oil had leaked [2, page 2]. This is a primary inspiration for this project.

Each day the National Oceanic and Atmospheric Administration (NOAA) published trajectory maps giving estimates of where the oil was located on the surface of the water, as well as its approximate density. These maps are archived online [9] with a guide to reading and interpreting them [11]. The maps themselves do not have a quantitative guide to precisely determine the amount of oil. However, they have a key to visually estimate the amount of surface coverage. The coverage varied considerably from day to day due to currents in the Gulf and weather patterns [1].

The National Aeronautics and Space Administration (NASA) also collected satellite images of coverage, but most of these images are difficult to parse visually due to clouds and the angle of the sun [1, 7]. Thus it is difficult for students to use them to form a dataset of daily estimates of surface oil. NASA worked on a spectral analysis of these images [6].

2.3 Institutional Context. Roosevelt University, with campuses in Chicago, IL, and Schaumburg, IL, is a private university with both undergraduate and graduate programs. The Roosevelt curriculum is based on principles of social justice. We used this project in an integral calculus course in Fall 2010 as part of an initiative to include semester-long projects in integral calculus classes at Roosevelt; more about this initiative can be found elsewhere [3]. The students in the course learned antiderivatives in differential calculus, began using Riemann sums at the start of the term, and finished the term with sequences and series. The project parts relating to areas and differential equations occurred when the class was working on Riemann sums and applications of integration, respectively. The integral calculus class had about thirty students, consisting of biology, chemistry, mathematics, and actuarial science majors. Their backgrounds varied, with some coming fresh out of first-semester calculus and others with a significant gap in their calculus sequence. An embedded tutor helped during class time, and this was essential on days when students did project work. Only one section was offered that fall, and so all students taking integral calculus did this project.

2.4 Motivation for Developing the Module. The practice of students working on a project in second semester calculus came from the involvement of the Roosevelt mathematics department with the Science Education for New Civic Engagements and Responsibilities (SENCER) project. Faculty members were awarded a small grant to redesign integral calculus to include a project that would accomplish the broad goal of SENCER: to connect science and civic engagement by getting students engaged in "complex, contested, capacious, current, and unresolved public issues" [13]. The BP oil spill was a perfect match for SENCER's stated goals. As estimates for the extent of the spill changed regularly [4], the project provided an opportunity for students to do some independent investigation into the question of how much oil was actually spilled.

2.5 Social Justice Connection. There are several ways for students to make connections between mathematics and social justice through this project. By having students develop their own estimates for the amount of oil spilled, we help students see that they have some power in a situation where information is being suppressed. Students can create models based on publicly available data. Students can understand how estimates are made, the importance of having working guidelines, and speculate on how to detect when a company is being evasive in its estimates. This can drive home why regulations and guidelines are important. Beyond this, students can research the implications of the spill, including the environmental impact, the economic impact, the effect on the lives of those in the coastal region, and the effect on wildlife, as well as possible long-term harm. Now that some time has passed, students can see what different states have done with their settlements from the spill and discuss what leads to an optimal use of funds. As a guided first step towards this, specific prompts are given in Appendix B.3.

3 Instructor Preparation

3.1 Before Class. The instructor intending to use this module should become acquainted with the various maps involved. Parts 1 and 2 involve estimating the surface area of the oil spill, based on maps from the NOAA website [9], and considering the errors in those estimates. NOAA provides a guide to reading the maps [11]. There are two main types of maps: nearshore and offshore. The nearshore ones show a smaller area, close to the coast, where there are dense amounts of surface oil. The offshore ones show a much larger region where a fine layer of oil covers the ocean's surface. Each map has a scale which visually shows the density of the oil. The scale itself is not precise, and no exact numbers are given on the map keys. Furthermore, the keys are different for each map.

Instructors should bring printouts of maps and rulers for the groups to use in class. After class the data from Part 1 should be saved and then added to the data from Part 2. In Part 2, the students are asked to submit their best estimate answers to a Google survey form which can then be used to create a dataset of surface oil estimates for the class to use. Instructors will need to set up such a survey prior to assigning the project part.

In the optional library portion, students meet with a librarian who works with them to find information and decide what is a credible source. It is particularly useful to make the students and the librarian aware of both the NOAA and NASA websites. If this is not an option, students can read an article such as [4] for background.

Table 10.2 in Appendix A describes the typical amount of oil (in barrels) that enters the North American coastal waters and the ocean worldwide over the course of a year. Note that the worldwide total is about 3.5 times the amount in the 2010 Deepwater Horizon spill, and the North American coastal water total is about half that of the Deepwater Horizon spill. Although this information is not directly used in the project, it is here to help the instructor put things into context. Note that the errors on these numbers may be quite large; they may be off by as much as an order of magnitude [10, page 69]. Over time, as a better understanding of natural seepages occurs and as programs to curb pollution are implemented, these numbers may need to be revised.

3.2 During Class. For all parts of the module referred to below, detailed worksheet questions and handouts are provided in the appendices.

In Part 2, the two questions about the date for the student's map are used to check that the students computed the days past the start of the spill correctly. The scales on the maps vary from day to day. Asking for the scale of the map used allows the instructor to easily compute the total amount of oil by multiplying the amount in each region by an appropriate scaling factor and summing to check the student's work. If answers are recorded using an online survey in a spreadsheet, it makes calculating and checking student work simple. It also allows the instructor to quickly create a spreadsheet of data for the class to use. The final three questions are used to get early feedback about the project and to help set up groups for the latter parts. If there is no group work component for the project, these questions can be dropped.

We rotated group members in the optional library portion, which helped them to establish good group dynamics in their final project groups.

In Part 3, one of the main points of emphasis is the use of unspecified constants, such as A, in differential equations. We do at least one example of this in class, emphasizing the fact that constants are sometimes represented by letters. It helps to give an example where a solution involves an unresolved integral, such as $\int f(t)dt$; otherwise, the students may not consider it as a possible part of an answer. Students complete this part of the project as a group, and they can be encouraged to use a computer algebra system to double-check their answers.

In Part 4, students model the volume of oil using both the dataset that the class creates and Table 10.1 from Appendix A. The instructor will need to use the class answers to Part 2 to create spreadsheets based on the total area covered (sum of the answers from 5-7 in Part 2), the area if the scale is linear (12a in Part 2), the area if the scale is logarithmic (12b in Part 2), and Table 10.1. Table 10.1 describes the total amount of oil (in barrels) lost by BP from day 1 up through day t of the spill. This is calculated via a combination of a flow-rate estimate [5] adjusted for the oil retrieved by the Riser Insertion Tube Tool, the Choke Line, and the Lower Marine Riser Package Top Hat #4 in barrels [1]. The error on these is estimated to be about 10 percent. Note that the standard oil barrel is 42 U.S. gallons.

3.3 Grading Notes. Because of the dependencies between the parts, quick feedback is essential. We used simple grading rubrics to help with this, and we e-mailed scores and comments to all members of the groups. We talked with the whole class about any major issues that arose.

When we used this module, each piece of the project had a different weight, with earlier parts worth less and later parts worth more. This allowed students to correct their work for future parts without having us re-grade the pieces multiple times. Typically, grades on the final poster and paper were very high. We set Part 1 and the library portion at 5% each, Part 2 at 10%, Parts 3 and 4 at 15% each, and the final poster and paper at 20% each. We also had students write reflective journals throughout the semester for an additional 10%. We optimized the grading rubrics for quick turnaround so that students could correct errors and use previous work in future project parts.

In Part 2, a spreadsheet can easily calculate total area from their regional areas in order to check the students' summations. In the spreadsheet, one can check question four by taking the answer to three, subtracting 20, and adding in 0 for April, 30 for May, 61 for June, 91 for July, and 122 for August in their answer to two. One can check question twelve by taking

$$\#12a = (\#5 * \#9 + \#6 * \#10 + \#7 * \#11) * 0.12$$
$$\#12b = (\#5 * 2^{\#9} + \#6 * 2^{\#10} + \#7 * \min(2^{\#11}, 100)) * 0.01.$$

We graded the best estimates in five, six, and seven based on whether they were the correct order of magnitude which led to quick grading without the need for the instructor to recalculate every individual number. About 5-10% of the class estimated the surface area using units of length rather than area but this was easily caught using the order of magnitude estimates.

In the optional library part, we included a penalty for improper citation and a heavy one for plagiarism. Because we made the library portion a small percent of the final project grade, it was used as a warning that students needed to write in their own words. In the past, we have allowed students to resubmit their reference list.

Note that it is ideal for students to have a chance to correct their work from Part 4 as they move on to Part 5, so it is important to provide feedback on Part 4 in a timely manner.

Because in mathematics evaluating papers and student writing is less common than grading exams, we provide specific detail for a simple grading rubric which we used, as follows. (The assignment itself is available in Appendix B.6.) We graded the paper out of 25 points. The title page, which is worth 0.5 points, must include the title, authors, and date. The abstract and reflection are both worth 2 points. The introduction is worth 3 points, with 1 point for the mathematical rationale and 2 points for the background of the spill. In the background, students must explain what happened and how the spill was contained. In data collection, worth 2 points, students must describe the units, time frame, and data source. The difference between methodology and the discussion of the model is often blurred, but this is not a problem. In the methodology, worth 2 points, students must explain why they chose their model. In the discussion of the model, students should discuss the pros, worth 3 points, and the cons, also worth 3 points. Students must explain the benefits of their model for the pros and data unreliability and changes in future conditions in the cons. They must describe the relationship to the physical world, worth 3 points. They should include a description of what the model is used for, when it is valid, and any long-term behavior. Their graphical representation of the model and equations is worth 2 points. The bibliography is worth 2.5 points; students who do not include access dates for the websites cited earn at most 2.

4 The Module

The project consists of five parts. In our original implementation, we had a new part due every 2-3 weeks, but each of the first four pieces can be used independently with minor modification. The fifth piece stresses mathematical communication, uniting the previous parts. Copies of the project handouts are in the appendix. Below we offer a possible timeline to use the project in a fifteen-week semester.

| Week 1 | Assign project Part 1. Give students 30 minutes in class to work on it in groups, and assign the rest for homework. If assigning the optional library part, meet with a librarian to arrange a session for your class. Create an online survey to collect their answers to Part 2. |

Week 2	Collect Part 1. Give feedback on Part 1, either as general comments to the class or in the form of graded assignments. Students will use this feedback when completing Part 2. Assign Part 2.
Week 3	Collect Part 2.
Week 4 & 5	Students do library research on the context of the oil spill. Alternatively, instructors can provide them with an article or two to read, e.g., [4]. Give feedback on Part 2.
Week 6	(after covering techniques of integration and introductory differential equations) Assign Part 3 to groups of about four students. Give them some class time (20-30 minutes) to work on it in order to facilitate group collaboration.
Week 7	Spend about 20 minutes on group work for Part 3 and help answer questions about the more difficult models.
Week 8	Collect Part 3.
Week 9	Give feedback to students on Part 3. Assign Part 4. Give the class the data set from Part 2. Set aside some class time (1 hour) for students to work in groups on curve fitting on computers.
Week 10	Collect Part 4.
Week 11	Give feedback on Part 4. Assign Part 5. Give groups 20 minutes in class to coordinate.
Week 12	Draft of posters due; print copies for groups to peer-review in class (20-30 min).
Week 13	Part 5 due.
Week 14	Students present posters/submit group paper.
Week 15	Include one or two questions on the final exam related to the project.

Part 1 of the project is on estimating areas, sums, and maps; see the handout in Appendix B.1. For this part of the activity, we pair up the students and provide them with instructions, a ruler, and a map with each pair having a different map. After 20-30 minutes, the pairs meet with another pair and explain the rationale for their area estimates. Students need to use their own judgment to decide how to apply the oil density scale provided, either by assigning one box to each color or averaging values for multiple boxes. Students decide for themselves how best to estimate the areas; there is no fixed right answer. Students are asked to determine upper and lower bounds for their estimate and to calculate an error. This foreshadows Riemann sums and upper and lower integral estimates in a context where they are needed. Students need to articulate their reasoning in the discussion of their answers.

After class, we assign Part 2 as homework. In this part, each individual student is given a specific date after the spill. For that date, they are instructed to complete a list of questions about their specific set-up and enter their answers into an online survey to create a class-wide dataset. See Appendix B.2 for the accompanying handout. Students need to find and use the corresponding map to create estimates. They typically use triangles and rectangles to create these area estimates. Our students have the most difficulty expressing areas using sums.

At this point students are ready for the optional library component; see Appendix B.3 for the accompanying handout. In this part, students conduct library research on the context of the oil spill and learn about how to find information and decide what is a credible source. Students respond to prompts such as, who was collecting the data; how were the maps being made; how are people affected by the spill; and, how was the economy affected? Alternatively, instructors can provide an article or two to read, e.g., [4]. After the reading assignment instructors can lead a class discussion on these same questions (who was collecting the data; how were the maps being made; how are people affected by the spill; and, how was the economy affected?), thus situating the mathematical work they are doing in its broader social context.

After covering techniques of integration and introductory differential equations, we assign Part 3 to groups of about four students. We give them some class time (20-30 minutes) to work on it in order to facilitate group collaboration. Part 3 is independent of Parts 1 and 2. Here, students work toward developing a model using differential equations; we include an accompanying handout in Appendix B.4. The scenarios in Part 3 involve the use of constants in differential equations, a skill that is essential for creating models, but is often overlooked in the calculus curriculum; it is good to do an example in class. For students who are unfamiliar with some of the techniques, such as solving linear differential equations, pieces can be omitted or the use of technology encouraged.

Students must make connections to what is physically reasonable in this particular setting in Part 4. This part depends on both Part 1 and Part 2, but it can be shortened to be independent of these. This section involves curve fitting and deciding when models make sense. See Appendix B.5 for the corresponding handout. We give the class the data set from Part 2 and set aside some class time (1 hour) for students to work in groups on curve fitting. The surface data from Part 2 vary wildly; it is difficult to fit them into a model well. The overall data from Table 10.1 work better. Students need to consider where the modeling function should be increasing or decreasing, what its concavity is, and whether to use a piece-wise function in their model. When we use this module, we have students spend some class time at a computer lab in order to orient them to the technology (depending on instructor preference, this can involve Maple, Mathematica, Matlab, or, for simpler fits, Excel).

Here, the emphasis is on justifying why one model should be chosen over another. Students also need to think about what piece-wise functions represent and how they can be used as models. Note that Table 10.1 is linear in the beginning, and then it incorporates the oil that was recovered. Thus, a piece-wise model is realistic. When used in class, the student groups will come up with different models from one another, and they should be able to provide some justification for their answers.

For Part 5, students are asked to create a poster or write a research paper; see Appendix B.6 for the directions we use for this assignment. We instruct students to write for an audience who has taken calculus. Having students peer-review a draft of the posters leads to a significant increase in quality. When we do this and other projects, we have a poster day where students can see other groups' work and ask questions of other students. In most semesters, our students presented their posters at a university-wide STEM (Science, Technology, Engineering, and Mathematics) event. If such an event does not occur, students can still benefit from presenting their work to the class.

This final part is done in groups, and a poster session can be held for students to ask other groups about their choices. This leads to a rich discussion of why one model is better than another, the difficulties in finding precise data, and the limitations of modeling.

5 Additional Thoughts

5.1 Application Notes. When we implemented this module, students worked in groups of four for all of the project parts except for Part 2. Each individual student was tested on the content in each of the project parts, either by having them do their own area estimate for a map or by including an exam question that was similar to a small piece of a part. This helped students feel that every individual was accountable for every part of the whole. We explicitly told students this would be the case, and it seemed to help the group dynamics.

By the end of the semester students came to enjoy the project, although they thought it was difficult while they were doing it. At the end of the semester, students were invited to respond to an online survey of the course. Only three did, but they commented favorably on the project; here is one such comment: "After using differential equations to solve real world problems such as the Gulf spill, it got me thinking that math can be used in a lot of different fields". A different student remarked, "I came to understand what the point of calculus is and why it is applicable to everyday things. I had really no idea during Calculus I". As students did all of the standard application problems in differential calculus, this shows the usefulness of a project that uses actual data and situations.

5.2 Possible Variants. In a class where students have not done Parts 1 and 2, one could modify Part 4 so that students are given the data in Table 10.1 to use, and most of question 1 could be eliminated. Alternatively, students in a class with more time to dedicate to a project could be given references to generate Table 10.1 on their own.

The spill lends itself to other modeling projects that consider the cost (in oil, livelihood, wildlife, etc.) which could be done in a follow-up math-modeling course. Similarly, one could see how estimates of the spill correlate to gas prices. As a variant of the paper, students could argue in writing for the cost of the spill, either from the government's perspective or from BP's perspective. There is a wealth of information archived at the NOAA website [8]. The project could be expanded to include more detailed models that take droplet size into account while calculating how oil disperses and dissolves [2, pages 12, 22] or into a system of mass-balance equations for the oil inputs and outputs [2, Appendix I]. These equations tie into probability and statistics, optimization problems, and volumes.

Appendix A: Tables and Datasets

Table 10.1. Cumulative oil (in barrels) that entered the Gulf based on flow-rate estimates [5], adjusting for oil recovered by BP [1]

Day	Total Oil	Day	Total Oil	Day	Total Oil	Day	Total Oil
1	5000	23	1343013	45	2616605	67	3459022
2	67000	24	1402538	46	2664639	68	3491212
3	128888	25	1461950	47	2708787	69	3521418
4	190663	26	1521250	48	2752609	70	3553606
5	252325	27	1580438	49	2795460	71	3583557
6	313875	28	1639513	50	2838563	72	3613208
7	375313	29	1698475	51	2881351	73	3642791
8	436638	30	1757325	52	2924493	74	3672451
9	497850	31	1816063	53	2967304	75	3701926
10	558950	32	1874688	54	3009741	76	3731460
11	619938	33	1933200	55	3056988	77	3761015
12	680813	34	1991600	56	3096295	78	3790587
13	741575	35	2049888	57	3128371	79	3819604
14	802225	36	2108063	58	3161029	80	3858052
15	862763	37	2166125	59	3197035	81	3903303
16	923188	38	2224075	60	3230630	82	3948324
17	983500	39	2281913	61	3261537	83	3984423
18	1043700	40	2339638	62	3291002	84	4024581
19	1103788	41	2397250	63	3330535	85	4068111
20	1163763	42	2454750	64	3363039		
21	1223625	43	2514372	65	3394567		
22	1283375	44	2567787	66	3427724		

Table 10.2. Oil (in barrels) that enters North American coastal waters and ocean waters in a typical year (1990-1999) prepared with data from [10, page 69]. Oil extraction and transportation includes platform spills, produced waters, spills from tanker/barges, pipelines, coastal facilities, and atmospheric deposition. Oil consumption includes river and urban runoff, oil spills from cargo ships, operational discharges from commercial and recreational vessels, and atmospheric deposition.

Source	North American Costal Waters	Worldwide
Natural Seeps	1,120,000	4,200,000
Oil Extraction and Transportation	169,000	2,656,000
Oil Consumption	1,173,000	6,696,000
Total	2,462,000	13,552,000

Appendix B: Assignments and Handouts

B.1 Project Part 1: Estimating Areas, Sums, and Maps. THIS IS A WORK-SHEET STUDENTS USE IN CLASS.

(1) Using the scale on the perimeter of the picture, create a rough estimate of the area that is covered with oil by doing the following

(a) Label the region on the picture that is covered with oil.
(b) Estimate the largest area that might be covered by oil.
(c) Estimate the smallest area that must be covered with oil.
(d) Make your best estimate for the area covered with oil.
(e) Give a range for the likely error associated with your estimate.

(2) What assumptions did you make when deciding which region of the picture contained the spill?

(3) What assumptions did you make when estimating the error?

(4) Your map has a scale that shows how the shades of blue correspond to the density of oil on the water's surface. The scale shows how these blue shades correspond to eight squares, where each square has a different amount of black (oil) in it. Together with your group members estimate the total area of the surface covered with oil if the squares in the key are

(a) covered with a percent of oil according to a linear scale,
(b) covered with a percent of oil according to a logarithmic scale.

Scale	Box 1	Box 2	Box 3	Box 4	Box 5	Box 6	Box 7	Box 8
Linear	12%	24%	36%	48%	60%	72%	84%	96%
Logarithmic	1%	2%	4%	8%	16%	32%	64%	100%

(5) What assumptions did you make when deciding how to apply the scale to estimate the area of the surface covered by the spill?

(6) What assumptions did you make when estimating the error?

(7) Suppose the map has an area of a_k covered with the percent of oil represented in square k (k=1,2,...,8) in the map key. Write a formula using a summation for the total area of the surface covered with oil if the squares in the key are

(a) covered with a percent of oil according to a linear scale,
(b) covered with a percent of oil according to a logarithmic scale.

(8) Where might such a formula be used?

B.2 Project Part 2: Creating a Class-Wide Dataset. BELOW IS A LIST OF QUESTIONS WE ASSIGN AS HOMEWORK AFTER STUDENTS COMPLETE PART 1. IF USED IN A PRECALCULUS OR QUANTITATIVE LITERACY COURSE, QUESTIONS 7 AND 8 MAY BE OMITTED.

(1) What is your name?

(2) What is the month for your data?

(3) What day of the month corresponds to your data?

(4) The well exploded on April 20, 2010. How many days after that is the date for your map?

(5) For the near-shore estimate, what is the total area with a light covering? Give your answer in square miles.

(6) For the near-shore estimate, what is the total area with a medium covering? Give your answer in square miles.

(7) For the near-shore estimate, what is the total area with a heavy covering? Give your answer in square miles.

(8) For the near-shore estimate, what is the total area that is listed as uncertain? Give your answer in square miles.

(9) On the distribution scale at the bottom of the map, what box best corresponds to the light-blue region?

(10) On the distribution scale at the bottom of the map, what box best corresponds to the medium-blue region?

(11) On the distribution scale at the bottom of the map, what box best corresponds to the dark-blue region?

(12) Your map has a scale that shows how the shades of blue correspond to the density of oil on the water's surface. The scale shows how these blue shades correspond to eight squares, where each square has a different amount of black (oil) in it. Estimate the total area of the ocean's surface covered with oil if the squares in the key are

(a) covered with a percent of oil according to a linear scale,

(b) covered with a percent of oil according to a logarithmic scale.

Scale	Box 1	Box 2	Box 3	Box 4	Box 5	Box 6	Box 7	Box 8
Linear	12%	24%	36%	48%	60%	72%	84%	96%
Logarithmic	1%	2%	4%	8%	16%	32%	64%	100%

(13) Please give any comments/concerns you have about the data.

(14) For your final group, how do you prefer to meet? List days and times that you are available in person or whether you prefer online meetings.

(15) Please state any other comments/concerns you have about groups.

B.3 Optional Project Part: Library Research. THIS IS AN OPTIONAL COMPO-
NENT OF THE PROJECT. STUDENTS WORK ON THE FOLLOWING QUESTIONS INDIVID-
UALLY, WITH THE HELP OF A LIBRARIAN OR WITH INSTRUCTOR-PROVIDED REFER-
ENCES.

We want to find out more about the background of the Deepwater Horizon oil spill. Using the library, the internet, and credible sources, answer the following in your own words in complete sentences. Include references for all sources used.

(1) How did the Deepwater Horizon spill occur?

(2) Cleaning up the spill:

(a) When did the different clean up strategies begin?

(b) What was involved in them?

(c) How long did each strategy last?

(d) For each strategy, what was the outcome of the strategy? Explain whether it helped, and if so, by how much.

(3) Data for the spill

(a) Who is collecting data on the spill?

(b) How is the amount of spill estimated?

(c) How were the maps of the spill areas created?

(d) What sources of uncertainty were involved, and how did the size of the uncertainties compare to the size of the estimates?

(e) Why is an accurate assessment useful?

(4) What were the impacts of the spill on the coastal region?

(a) How were people affected?

(b) How was the environment affected?

(c) How was the economy affected?

(d) What responsibilities did the corporations involved incur?

(e) How did states use their settlements from the spill? Were there behaviors that led to a more effective use of funds?

B.4 Project Part 3: Developing a Model using Differential Equations.

THIS IS A WORKSHEET STUDENTS USE IN CLASS.

(1) We want to create a differential equation that models the oil spill as accurately as possible. To do this, we'll consider a few different cases. Write a differential equation for each of the following scenarios. For each case, clearly explain what the variables are and what they represent, as well as what constants are involved. (Note that we won't know specific values of the constants yet! They will be things like A or k.)

(a) Oil is being added at a constant rate A.
Sample Answer: y is the amount of oil, t is the time since the start of the spill. The differential equation is $\frac{dy}{dt} = A$.

(b) Oil is being added at a constant rate A, and it is being removed at a constant rate B.

(c) Oil is being added at a constant rate A, and it is being removed at a rate that is proportional to the amount of oil in the water.

(d) Oil is being added at a constant rate A, and it is being removed by two different mechanisms. One removes it at a rate that is proportional to the amount of oil in the water. The other removes it at a constant rate B.

(e) Oil is being added at a constant rate A, and it is being removed at a rate that is a function of time, $f(t)$.

(f) Oil is being added at a constant rate A, and it is being removed by two different mechanisms. One removes it at a rate that is a function of time, $f(t)$. The other removes it at a constant rate B.

(g) Oil is being added at a constant rate A, and it is being removed by two different mechanisms. One removes it at a rate that is proportional to the amount of oil in the water. The other removes it at a rate that is a function of time, $f(t)$.

(h) Oil is being added at a linear rate $At + B$, and it is being removed at a constant rate C.

(i) Oil is being added at a linear rate $At + B$, and it is being removed at a rate that is proportional to the amount of oil in the water.

(j) Oil is being added at a linear rate $At + B$, and it is being removed at a rate that is a function of time, $f(t)$.

(k) Come up with a (reasonable) scenario of your own! Explain the scenario and the differential equation involved.

(2) For each differential equation, describe what physical scenario it can correspond to.

Sample Answer: For part (a), we have a broken pipe with oil leaking at a constant rate, but there is nothing being done to clean it up.

(3) Solve each differential equation in (a)-(d), (h), (i), and (k); your answers will involve constants. These solutions are different types of models that could apply to our situation.

Sample Answer: For part (a), we have $\int dy = \int A dt$, so $y = At + C$ where C is a constant.

(4) Solve each differential equation in (e)-(g) and (j) for the following choices of $f(t)$ where D is a constant. Your answers will involve constants.

(a) What are the solutions if $f(t) = Dt$?

(b) What are the solutions if $f(t) = Dt^2$?

(c) What are the solutions if $f(t)$ isn't specified? Your answer to this will involve $f(t)$.

(5) Which of these scenarios seem reasonable to you? For at least one scenario describe (with references) why it does not accurately model the events in the Gulf. For at least one scenario describe (with references) why it might accurately model the events in the Gulf. (In our next part, we will combine these scenarios with data!)

B.5 Project Part 4: Model Creation. THIS IS A WORKSHEET FOR GROUP WORK; STUDENTS TYPICALLY START WORKING ON IT DURING CLASS.

(1) Graphs of Various Data Sets

(a) Create and label graphs for each of the following.

(i) Accumulated oil from Table 10.1, which gives the cumulative barrels of oil spilled by day.

(ii) Surface Coverage for total area covered by light, medium, or heavy amounts of oil versus time.

(iii) Surface Coverage Linear for the area covered assuming a linear scale versus time.

(iv) Surface Coverage Logarithmic for the area covered assuming a log scale versus time.

(b) Describe what may have actually occurred physically that would explain why the class surface coverage data looks the way it does.

(c) How reliable do you think the class data is and why? Can real-life data ever be 100% reliable? What could lead to an underestimate or overestimate of the oil?

(d) Which of these graphs looks easiest to fit a curve to? Explain why.

(e) Which of these graphs looks hardest to fit a curve to? Explain why.

(2) Models of Data

(a) Use the easiest-to-fit data from question 1d to create a model of the spill based on the solutions to Project Part 3. If you use the surface coverage data, you may assume that the actual amount of oil is proportional to the surface amount.

    (i) Write the equation that best fits your data.

    (ii) Indicate what is represented by each of the quantities in your model.

(b) Some models are piece-wise defined. This means they exhibit one type of behavior during one time interval, and a different behavior during a second interval. Do any of the data sets have a possible transition point from one model to another?

    (i) For each portion of data, what type of model looks most appropriate? Explain (mathematically) why.

    (ii) Write out the piece-wise formula for this data set.

    (iii) Describe physical conditions that may have led to this situation.

(3) How Well the Data Fits the Model

(a) Determine which of the models you tried is the best fit. Explain why (mathematically) it is a good fit, and what (physically) that means.

(b) Determine which of the models you tried is the worst fit. Explain why (mathematically) it is a bad fit, and what (physically) that means.

(c) Using your best model, find an estimate of the amount of oil on the surface of the Gulf on August 1, 2010.

(d) Graph your best-fit model and the corresponding data on the same plot. Discuss how well your model fits the data.

(4) Derivatives and Net Daily Changes

(a) Calculate the derivative of your best-fit model. Give an equation for this.

(b) Graph the derivative of the curve for your model and the data for the net change in oil on the same plot. Does your model fit the data you gathered?

(c) How are the net daily change in oil and the derivative related mathematically?

B.6 Final Project Part: Paper/Poster Creation. THE DIRECTIONS FOR THE FINAL PART OF THE PROJECT ARE PROVIDED BELOW. IN THIS PART STUDENTS ARE ASKED TO CREATE A POSTER OR WRITE A RESEARCH PAPER.

The final part of the project is to compile your findings into a poster presentation and/or research paper. The required elements are as follows:

(1) Identifying information

(a) Title of project

(b) Names of group members

(c) University name

(d) Date

(2) Introduction and background on the Deepwater Horizon oil spill and the resulting spread of oil

(3) Description and explanation of data collection, data sources, and methods of handling data

(4) Summary and tables of data used in your model

(5) Presentation of your model with all terms explained

(6) Graphs of data and model

(7) Discussion of results

(a) Strengths and advantages of your model

(b) Limitations and problems with your model

(8) References for any sources used

Bibliography

[1] Department of Energy. *Data from Deepwater Horizon*. Available online at http://www.energy.gov/about-us/open-government/data-deepwater-horizon, accessed on September 1, 2016.

[2] The Federal Interagency Solutions Group, Oil Budget Calculator Science and Engineering Team, *Oil Budget Calculator: Deepwater Horizon-Technical Documentation*, A Report to the National Incident Command, November 2010.

[3] B. González-Arévalo and M. Pivarski, "The Real-World Connection: Incorporating Semester-Long Projects into Calculus II," *Science Education and Civic Engagement: An International Journal*, Volume **5** Issue 1 (2013), pages 17-24.

[4] Ian R. MacDonald, "Deepwater disaster: how the oil spill estimates got it wrong," *Significance*, Volume **7** Number 4 (2010), pages 149-154.

[5] McNutt et. al., "Review of flow rate estimates of the Deepwater Horizon oil spill," *PNAS*, Volume **109** Number 50 (2012), pages 20260-20267.

[6] National Aeronautics and Space Administration. *Gulf of Mexico Initiative Targets Oil Spills and Other Ecological Challenges*. Available online at https://www.nasa.gov/topics/earth/features/oilspill/oilspill-calipso-caliop.html, accessed on September 1, 2016.

[7] National Aeronautics and Space Administration. *Oil Spill Features*. Available online at https://www.nasa.gov/topics/earth/features/oilspill/, accessed on September 1, 2016.

[8] National Oceanic and Atmospheric Administration. *Deepwater Horizon Oil Spill*. Available online at http://response.restoration.noaa.gov/deepwater-horizon-oil-spill, accessed on September 1, 2016.

[9] National Oceanic and Atmospheric Administration. *Deepwater Horizon Trajectory Maps: By Date*. Available online at http://response.restoration.noaa.gov/oil-and-chemical-spills/oil-spills/response-tools/deepwater-horizon-trajectory-maps-dates.html, accessed on September 1, 2016.

[10] National Research Council (NRC) of the National Academies of Science. *Oil in the Sea III: Inputs, Fates, and Effects*, 2003.

[11] Office of Response and Restoration, NOAA's National Ocean Service *Interpreting NOAA's Trajectory Prediction Maps for the Deepwater Horizon Oil Spill*, National Oceanic and Atmospheric Administration. Available online at http://response.restoration.noaa.gov/sites/default/files/NOAATrajectoryMaps.pdf, accessed on September 1, 2016.

[12] Jonathan L. Ramseur, *Deepwater Horizon Oil Spill: The Fate of the Oil*, December 16, 2010 / January 5, 2011, Congressional Research Service Report for Congress 7-5700 R41531. Available online at `https://www.fas.org/sgp/crs/misc/R41531.pdf`, accessed on September 1, 2016.

[13] SENCER-National Center for Science and Civic Engagement. *The SENCER Ideals.* Available online at `http://sencer.net/sencer-ideals/`, accessed on July 10, 2018.

11

Voting with Partially-Ordered Preferences

John Cullinan and
Samuel Hsiao

Abstract. This module introduces a method – the partial Borda count – for voting with partially-ordered preferences. The partial Borda count allows voters to vote with their true preferences rather than having a linear preference order imposed on the ballot structure a priori. Furthermore, this method provides a simple and mathematically consistent way of scoring truncated ballots and improves upon so-called *bullet voting* and other methods of resolving incomplete ballots. The mathematical content of the module is elementary combinatorics with applications to social choice theory. We present the material for the classroom following an inquiry-based format with detailed examples and notes for the instructor. The take-home message for the student is threefold: there are inherent social problems with requiring linearly-ranked ballots; there is a simple procedure for allowing an arbitrary preference ranking to be expressed; and that method is the most desirable among all possible procedures for voting with partially-ordered preferences. **Keywords:** Social choice theory, voting preferences, Borda count, partially-ordered sets, combinatorics, inquiry-based learning, problem-based learning, constructivist instruction.

The mathematical theory of voting with three or more alternatives has a rich history, going back at least to the mid 18th century. Most of the well-known results in this subject are based on the assumption that the voters have linearly ranked the alternatives. But what if a voter's true preferences involve ties or more complicated partial orders?

The goal of this module is to explore a method of voting with non-linear preferences, as well as some of the method's implications for social justice.[1] The mathematical techniques are self-contained and, depending on the level of the course, completely elementary. In particular, we introduce partial orders and use Hasse diagrams to represent ballots. The combinatorial properties of the posets then have implications for fairness properties of the partial Borda count voting system. From the point of view of social justice, the main contribution is to show that it is possible to incorporate arbitrary preference rankings into a ballot that satisfies all standard fairness axioms; that is, if one's true preferences are non-linear, then that voter would not be forced to submit a linearly ranked (and therefore disingenuous) ballot. The partial Borda count voting system also produces a simple, mathematically consistent way to score a truncated ballot. All of these topics will be covered in detail in the following sections.

1 Mathematical Content

Before beginning an increasingly detailed description of the mathematics, we will address three preliminary questions:

(1) **What courses would this module work for?**

   This module is designed for an introductory class on voting theory or, more broadly, a general education or mathematics for liberal arts course as an activity on the mathematics of voting. It fits naturally into a semester-long course or into a survey course in which voting theory makes up one section or learning module.

(2) **What are the prerequisite mathematical ideas?**

   The prerequisites for the students are modest. They should have familiarity with basic high school algebra and the instructor should review the following concepts:

   - Examples of Social Choice Functions (voting procedures), especially the Borda Count;
   - Fairness Criteria in Social Choice Theory.

   The purpose of this module is twofold: to introduce the notion of voting with partially-ordered preferences via the partial Borda count, and to show that the partial Borda count is, in a concrete sense, the mathematically "fairest" procedure for aggregating the ballots.

(3) **What will be the mathematical value of the module?**

   This module will strengthen the students' techniques of mathematical reasoning and introduce basic combinatorial principles, such as Hasse diagrams and real-valued functions on posets. While these topics seem esoteric for an introductory class, one of the purposes of this module is to demonstrate that through the lens of voting theory, they are actually quite concrete and have real-world significance.

---

[1] EDITOR'S NOTE: See Chapter 16 (*What Does* Fair *Mean?* by Kira Hamman) for another module involving voting theory.

The specific mathematical content that we believe to be new for the student is reviewed in Appendix A.1 (the new-for-instructor content is summarized in Section 3). In this section we are more concerned with the pure mathematics than with the applications to social choice theory. Therefore, we postpone definitions of the relevant fairness axioms until Section 4, and a review worksheet for social choice theory is provided in Appendix A.2.

This material is meant to be developed in the context of a class that focuses on voting theory and social choice theory and assumes no mathematical prerequisites. In fact, the course we use this module in has only modest prerequisites (a passing score on our institution's math diagnostic test, i.e., proficiency in high school algebra). Individual instructors can choose, depending on the level of the students, how much mathematical formalism to include. In our course, after introducing standard voting techniques for three or more alternatives over several classes, we spend a day on voting with partially-ordered preferences as a way to generalize linear preferences. The techniques and equations are intuitive and work well in both standard and inquiry-based formats.

By the time students attempt the module, they should be familiar with the basics of social choice theory. In particular, students should have worked on examples or homework involving the Borda count and some other voting methods for comparison (Instant Runoff Voting, Plurality, among others). They should also have seen some of the fairness criteria used to assess social welfare functions, such as Pareto, the Condorcet Winner Criterion, and Independence of Irrelevant Alternatives [8]. Ideally, this module would come at the end of the social choice theory section of the course. Instructors should be familiar with social choice theory and basic set theory/combinatorics. In particular, the instructor will need to explain the basic properties of (Hasse diagrams of) posets and adapt the fairness criteria usually reserved for linearly ordered ballots to partially-ordered ones.

2 Context / Background

2.1 Institutional Context. This module has been used by the first author at Bard College during a course on mathematics and politics – a general-education course with no mathematical prerequisites. Bard College is a small liberal-arts college with approximately 2000 undergraduate students. The primary mission of the college is undergraduate education and, with the exception of a handful of Masters programs, is solely concerned with the teaching of undergraduate students. Historically, Bard students have tended to major more in the arts and literature than in science and mathematics. However, one of the tenets of a liberal-arts education is breadth of knowledge, and there has always been a general-education mathematics requirement.

While some students use Precalculus, Calculus, and beyond to satisfy the mathematics requirement, there are students at the college who lack the background for these courses. Rather than teach general survey courses, each mathematics instructor at Bard has designed their own semester-long general-education course based on a single theme for non-majors (some examples include *Explorations in Number Theory*, *Chance*, *Secret Codes*, and *Mathematics and Politics*). This is the context in which this module was created.

Introductory courses on voting theory exist in many colleges and universities throughout the United States and, in addition, many general mathematical survey courses have voting theory components to them as well. The course at Bard College typically runs with 25-30 students.

The project to design a voting procedure that allowed for partially-ordered preferences began as a year-long senior thesis at Bard College. We then collaborated to generalize those results, in particular the uniqueness theorems, and the final product is [5]. The first author of this module uses this material in the classroom, much in the way it is presented below.

In 2013, we collaborated with a Masters student in Curatorial Studies at Bard College to use the partial Borda count to determine which art would be displayed in a curatorial exhibit. Over 200 people across campus (students, faculty, and staff) were interviewed about their art preferences and asked to rank, using partial orders if applicable, their true feelings on six aspects of an art installation. We then aggregated the ballots using the partial Borda count to help determine which aspects of the collection would be most visible to the public.

2.2 Underlying Social Justice Context. Almost all of the standard voting procedures (e.g., the Borda Count, Instant Runoff Voting, etc.) assume that a voter is able to rank alternatives linearly. But what if one's true feelings are more complicated – tree-like or non-linear? What if one is apathetic? Then the most-used voting procedures force voters to submit disingenuous ballots. Thus, we are interested in whether it is possible to create a new procedure, subject to standard fairness criteria in social choice theory, that allow for partially-ordered preferences. In particular, the fairness criteria should specialize to standard ones if all voters submit linearly ordered preferences.

A voting tactic known as *bullet voting* [7] is when a voter ranks only a subset of the alternatives and submits an incomplete ballot. (In fact, bullet voting specifically refers to the practice of voting for one alternative instead of a ranked ballot.) This practice is widely seen as a way to manipulate election outcomes because bulleted ballots carry more weight for the single alternative than in a fully complete ranked ballot. The partial Borda count provides a way to renormalize a bulleted ballot and score it with the others in a mathematically consistent way.

This module contributes to a conversation on social justice because it introduces a voting paradigm where voter preferences can be represented more accurately than in many of the other well-known methods. Our underlying assumption is that voting methods that represent voter preferences as truthfully as possible will make democracies more representative, more egalitarian, and fundamentally more just.

3 Instructor Preparation

Voting with partially-ordered preferences has been mentioned in the literature going back at least to Arrow's seminal work [2] with generalizations in [3] and [6]. The introduction of [5] gives some historical background and references and [1] serves as a gentle introduction to voting with partially-ordered preferences. In addition, Young's paper [10] (on which [5] was based) introduces a few different fairness criteria that are not typically discussed in introductory textbooks. Finally, we mention [4] for an encyclopedic reference on the many examples of social choice functions for partially-ordered profiles and the fairness axioms they satisfy.

We have written the teaching module with the idea that the class has covered the basic definitions of voting theory, whether as part of a semester-long course or as part of a longer module in the middle of a survey course. We have included a handout in Appendix A that we use in our course and give to the students just prior to the beginning of the section of partial orders. In that handout we use colloquial language to complement the class text [8] and we find it is well received by students as a study guide. For this chapter, we assume the instructor is familiar with the usual fairness criteria used to evaluate social welfare functions (e.g., Pareto, etc.).

The module will introduce posets and their Hasse diagrams. While we do not use much of the theory of partially-ordered sets (we use posets and their Hasse diagrams interchangeably), we recommend that the instructor reviews their basic properties. We recommend [8] for the background on social choice theory and [9] for the combinatorics and posets.

4 The Module

We assume that this module comes during a section on social choice theory, whether as part of an entire semester course in mathematics and politics, or simply a section in a survey course. As presented, the teaching module is likely too long for a one-hour class but could easily fit into two such classes. Alternatively, the material could be amended to fit into a single class of an hour and twenty minutes (the length of a typical class at Bard College). Our approach is primarily inquiry-based and problem-based, though is easily adaptable to more traditional teaching formats. As stated in Section 1, we assume that the audience (students) is familiar with the basic definitions and language of social choice theory. See Appendix A.2 for a brief overview of prerequisite material in social choice theory; the overview could easily be used as a handout in class for the students.

We have written this module with the intent that it take place over two 80-minute class periods, comprising a typical week of classroom time at the authors' institution. Before beginning in detail, we give a very quick overview of the module and what we will focus on in this section.

(1) **Review of the Borda count and approval voting.** We suggest beginning with a quick review of the Borda count and introducing the notion of poset by recalling the definition of approval voting and showing how approval ballots can be visualized via a Hasse diagram.

(2) **Inquiry-based approach to posets.** We will discuss specific examples below, but we prefer to use leading questions in class to have the students come up with the definition of poset and Hasse diagram. We find that because of the mathematical background of our typical audience, the students absorb the material better than if we start by defining posets on the board. Once the students are comfortable with the idea, we suggest asking them to generate interesting, fun examples from everyday life in order to further solidify the basic properties of partial orders.

(3) **The partial Borda count.** We will then show how we introduce the main topic of the module once the students have a good grasp of partial orders. Again, our approach is to let the students explore the material on their own with minimal direction from the instructor, though with well-chosen illustrative examples as guides.

Once several key examples have been discussed, we share our recommendations for explaining the scoring formulas for the partial Borda count.

(4) **Fairness.** Because fairness is a key issue for linearly-ranked ballots, it should be highlighted in the context of partially-ordered ballots as well. But how should we amend the standard fairness criteria to work for voting with partially-ordered preferences? That is the key question that we address in the final portion of the module.

4.1 Review of the Borda Count. We suggest beginning with a quick review of the Borda count. In order to make the partial Borda count which will be introduced agree with the classic Borda count when a linearly ordered ballot is submitted, we recommend scaling the Borda scoring by a factor of 2 (this will not change the outcome of an election or the fairness properties of the voting procedure). That is, given a ballot of m alternatives, we assign a score of $2m - 2$ to the most-preferred alternative, $2m - 4$ to the second most-preferred, etc., ending with a score of 0 for the least-preferred alternative. For example, given the ballot

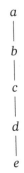

the alternatives $a, b, c, d,$ and e would receive 8, 6, 4, 2, and 0 points, respectively. Each ballot is scored according to this rule, and the alternative with the highest total score is declared the winner. It is important to point out to students that ties are possible, in which case the subset tied for the highest total score are all declared the social choices.

From the point of view of fairness, the Borda count satisfies the 'Always a Winner' criterion (since it always produces a non-empty subset of the winners as the social choice(s)) as well as Pareto and Monotonicity [**8**, p. 13]. However, the Borda count does not satisfy the Condorcet Winner Criterion (it may be the case that a Condorcet winner exists and the Borda count would select a different alternative as the winner) or Independence of Irrelevant Alternatives. To keep this exposition self-contained, we will not discuss these two particular fairness metrics further, but rather refer to [**8**] for more information.

4.2 Approval Voting. Next we suggest a quick review of approval voting. Below is a narrative replicating what we might say in our classes.

*A common method of voting when there are three or more alternatives is **approval voting** – for example, the American Mathematical Society uses approval voting for elections for the Nominating Committee and the Editorial Boards Committee [**11**, p. 1073]. Approval voting is the procedure whereby a voter partitions the alternatives into two subsets: those which are approved, and those which are* *not; no further refinements of the approved subset are given. Then, the alternative who is approved most often (or a subset tied for most approvals) is declared the social choice. This naturally leads to the idea of a partial order (details and definitions are given in Appendix A.1) and is a gentle introduction to voting with partially-ordered preferences. For example, given alternatives a, b, c, d, and e, suppose a voter approves of a and b only. Then those preferences can be represented graphically as*

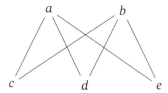

Indeed, any such ballot would be similarly partitioned, the approved subset lying above the unapproved, and edges indicating preference.

The main purpose of this module is to explore voting with arbitrary partial orders. In the next subsection we describe how instructors can introduce these ideas more formally in their classes. For more details and further development of approval voting, see [7].

4.3 Posets and Preferences. We find that when first introducing the notion of a poset, it helps to have students both generate their own examples and to come up with the notion of the Hasse diagram as a visual description of the partial on their own, with no preliminary definitions from the instructor. Below we list a few sample introductory questions that can be used:

(1) How would you graphically represent the following preferences among three ice cream flavors: *Chocolate is preferred to both vanilla and strawberry, but there is no preference between vanilla and strawberry.*

(2) Everyone in the class individually think of your own preferences. How would you represent them graphically?

For the first question we expect to see an answer such as Figure 11.1(a). For the second question, we might elicit a few responses from the students in the class. Some possible responses are presented in Figure 11.1(b)-(d).

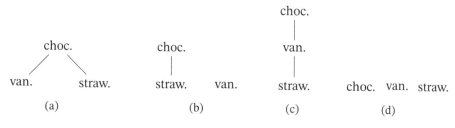

Figure 11.1. Some possible preferences among three flavors of ice cream.

While this may seem simple, we have found in practice that it takes several examples for an entire classroom to be comfortable with the notion of partially-ordered preferences, even for posets of size 3.

(3) What are some favorable properties of partial orders?

This is a hard question for the students since they have almost no exposure to the subject beforehand, but we are trying to motivate the students to come up with the following properties that characterize posets:

- An alternative cannot be preferred over itself;

- if a is preferred over b and b over c, then a is preferred over c.

It may also be instructive to point out the limitations of using posets to express preferences. For example, someone who loves all three flavors and has no preferences among them would use the same diagram, Figure 11.1(d), as someone who is merely apathetic. The *strength* of one's preference for any particular alternative is not directly encoded in a poset (nor is it in the traditional Borda count).

4.4 The Partial Borda Count. Who should win an election? This is the guiding question for this portion of the module and our goal is to have the students engage with it *before* giving them the formula for scoring the partially-ordered ballots. To start, we present a sample profile of ballots – perhaps continuing with the ice cream example – without giving information on how to score them, and ask students who they think should win. This makes students have to consider the role of antichains and singletons before giving away the formula from the partial Borda count and will make the formula more meaningful when it is presented. Specifically, we break this part of the class into the following steps.

Step 1. First, we give students a single poset and ask how they would determine the winner(s) if that were the only ballot. We do this with a few different types of posets of increasing complexity. For example, the following posets could make for interesting discussion:

The goal here is to get students thinking about how much weight to give alternatives based on their relative position to other alternatives. In the poset shown in the middle, for example, it is reasonable to say that a should win because it is preferred over more alternatives than c, even though a and c are at the same "level" in the poset.

Step 2. Next, we give an example of a profile of ballots and ask the students who the overall winner should be. We try to keep it simple enough so that it is fairly obvious who comes out on top on each ballot, but put a variety of ballots so it is not so clear who the overall winner should be, as in this example:

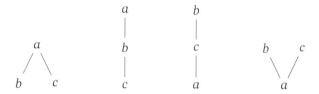

In this profile, it is clear that the first two ballots show preference for a over the others, the third ballot shows preference for b over the others, and the fourth ballot shows equal preference for b and c over a. If these were the only four ballots submitted, who should win? It may not be possible to come with a satisfactory resolution in class until the partial Borda count is presented. In this step our goal is just to get students thinking about what features of the posets should matter when aggregating ballots to determine the social choice. (It turns out that b is the winner if we score using the partial Borda count.)

Step 3. Third, we present the class with a single poset with three alternatives and tell them they have 6 points to distribute among the alternatives, with the more preferred alternatives getting more points. How should this be done? We then repeat with different types of posets.

For example, in a three-element antichain it should be clear that every alternative should get 2 points.

What about a chain? With a chain, say $a > b > c$, one would want to assign the most points to a, fewer points to b, and even fewer to c. There are several ways to do this, but if one wants to enforce the (reasonable) property that the point differential between a and b should be the same as that between b and c, there are just two possibilities: $(3, 2, 1)$ or $(4, 2, 0)$.

To decide, it helps to look at yet another example.

Consider the poset with relations $a > b$, $a > c$. (This is the first of the four posets shown in Step 2.) For this poset it is reasonable to assign equal points to b and c by symmetry; moreover the number of points assigned to a in this poset should arguably equal the number of points assigned to a in the chain $a > b > c$, because in both examples a is directly preferred over all the other alternatives. If a were given 3 points, there would be no way to equally divide up the remaining 3 points among b and c in the poset $a > b$, $a > c$.

By this reasoning the most plausible assignment of points to the chain $a > b > c$ would be $(4, 2, 0)$.

Step 4. Finally, we reveal equation (11.1) and discuss its interpretation in terms of "giving points away" to more preferred alternatives. We then use this equation to calculate weights and the social choice for the profile used in Step 2 and present additional examples as needed.

4.5 Fairness and Social Justice Implications. We believe that some discussion of the (mathematical) fairness properties are essential to the social justice component of the module, and so below we provide a very quick background that could easily be turned into a handout for the class.

Background on Fairness Criteria. Given a voting procedure, there are several standard metrics by which to evaluate it, often called **fairness properties**. **Arrow's Theorem** states that the only voting procedure that satisfies Pareto, Independence of Irrelevant Alternatives, and Monotonicity is a dictatorship (i.e., choose a ballot at random and select the highest-ranked alternative). Similarly, the **Gibbard-Satterthwaite Theorem** shows that a dictatorship is the only voting procedure for three or more alternatives that does not produce ties for the social choice, satisfies Pareto, and is non-manipulable. For more information see [8, Ch. 7].

Even though a dictatorship satisfies the standard fairness criteria, it is not a socially acceptable voting procedure. So, we necessarily must give up some notions of fairness in order to hold an election. In [10], Young proved that the Borda count is the unique voting procedure for linear ballots that is simultaneously **neutral**, **consistent**, **faithful**, and has the **cancellation property**. For precise definitions of these terms see [10] or [5], though for ease of exposition we give quick explanations:

- **Neutral.** Each alternative has an equal chance of winning before voting begins (i.e., if the order of the alternatives is permuted before voting, then the social choice(s) are permuted accordingly).

- **Consistent.** If disjoint sets of voters would choose the same social choice(s), then the union of the sets would as well.

- **Faithful.** If there is only one voter, and that voter prefers b over a, then a is not a winner.

- **Cancellation Property.** If, for all pairs of alternatives (a, b), the number of ballots where a is preferred to b equals the number where b is preferred to a, then a tie among all alternatives is declared.

It can further be shown that these properties imply the Borda count is **monotone** and satisfies Pareto, two well-known and important properties for voting procedures. Likewise, in [5] the partial Borda count is shown to be the unique social choice function that is consistent, faithful, neutral, and has the cancellation property. This generalizes Young's result to include profiles of partially-ordered ballots, with no restrictions on the partial order.

Suggestions for the Classroom. Our major goal for the classroom is now to address the following question: Why is the partial Borda count a reasonable voting system? Based on the prerequisites laid out in Section 1 and Appendix A.1, we feel comfortable posing the following questions to the class. For each fairness criterion, we have the students break into groups and try to answer the following questions about that criterion.

- How would you phrase this criterion so it makes sense for partially-ordered preferences?

- Does the partial Borda procedure satisfy this criteria?

In most standard introductions to mathematics and politics and voting theory, the concepts of monotonicity and Pareto are thoroughly developed and we tend to focus on these two concepts and use the following approach to explaining them in class. Again, we break the students into groups and have them work together to answer the following questions:

(1) How would you generalize the Pareto condition to partially-ordered ballots? What aspects of Pareto are the most salient, and what should be carried over to the context of partial orders? For partial orders, the Pareto condition carries over naturally:

Pareto condition: *For two alternatives a, b, if every voter prefers b over a, then a is not a winner (i.e., not in the social choice set).*

(2) What does monotonicity mean in the context of partial orders? This is more delicate, so we suggest beginning with examples where there are two alternatives with arbitrary partial orders, so the only possibilities are chains and antichains. For partial orders, the monotonicity condition must be altered slightly to accommodate incomparable elements.

Monotonicity condition: *Let p be a profile and let a be a social choice. Suppose one of the voters changes her original preference order from $\leqslant$ to a preference order $\leqslant'$ with the property that for all $b, c \in A - \{a\}$,*

$$b < c \Longleftrightarrow b <' c, \quad b < a \Longrightarrow b <' a, \quad \text{and} \quad a \not< b \Longrightarrow a \not<' b.$$

Then a remains a social choice for the new profile p'.

By focusing on these two criteria, already known to the students from linear ballots, the nuances of partial orders can be explored while always having the colloquial context of voting to explain the results.

Finally, we suggest showing that the partial Borda count does indeed satisfy Pareto and Monotonicity. At that point we simply give a brief overview of the further properties enjoyed by the partial Borda count (listed above). The uniqueness result is beyond the scope of this module, but we feel it is worth stating to students for completeness.

Social Justice Implications. Now that the students have worked through the mathematics and have a good internalized idea of partially-ordered preferences, we bring up the social justice implications for the partial Borda count. First, the partial Borda count allows for the most general types of preferences to be allowed on a ballot and has been mathematically proven to be the unique such voting method that satisfies the standard fairness axioms. That is, by allowing arbitrary partial orders to be submitted on ballots, we do not force the voters to submit disingenuous preference orders if their true feelings are non-linear.

Second, the partial Borda count allows us to improve upon the drawbacks of the tactic of *bullet voting* [7]. Bullet voting is what occurs when a voter is supposed to submit a linearly-ordered ballot, but only submits one alternative or a proper subset of alternatives. Do we score that ballot or discard it? If we discard a large number of bullet-ballots, then we are not accurately portraying the collective choice of the voters. More generally, what if a voter submits a partially complete linearly-ordered ballot? How do we evaluate these ballots along with those that are fully complete?

The partial Borda count provides a simple answer that scores the bulleted ballots along with the completed ballots in a mathematically consistent way: place all the unscored alternatives into a single antichain and then proceed via the partial Borda count. This way, it is not possible to use the bullet voting tactic to game the original voting system.

5 Additional Thoughts

5.1 Application Notes. The first author has taught this module five times in five iterations of a semester-long introduction to mathematics and politics. The material is extremely well received by the students, and they tend to find it a more fulfilling conclusion to social choice theory than Arrow's Theorem (and indeed, we have presented it in parallel with Arrow's Theorem). We recommend taking the appropriate time to allow the students to generate their own examples of posets. An initial time investment in this aspect of the class helps solidify their intuition once the equations are introduced.

5.2 Extending the Module. There is a significant increase in the level and technicality of the mathematics involved in so-called "fairness" properties of voting procedures, both for traditional linearly-ordered ballots and for partially-ordered ones as well. Because of the leap in sophistication and taking the intended audience's preparation into account, we omitted many of the details from that part of the module. The intent of this module is to comprise one or two classes of a semester and working through the details of the proofs is likely outside the scope of such a presentation. However it is conceivable that interested instructors could expand upon this section and explore fairness criteria in more depth.

5.3 Possible Variants. We have found this material to be amenable to student research projects at the sophomore and junior level. As an example, consider the following questions: Given an integral vector $(n_1, \ldots, n_k)$ such that $\sum_j n_j$ is even, 1) do the n_j represent the weights of a partially-ordered ballot and 2) if so, is there a unique such ballot? The answer to both questions is no, and we have worked with students to characterize the ballots that describe these surjectivity and injectivity questions.

Finally, even though the material has been presented for non-majors, we believe the module could be easily adapted for an upper-level course on combinatorics, possibly for a capstone course. The article [5] is self-contained and could be read by advanced students.

Appendix A: Assignments and Handouts

A.1 Overview of the Mathematical Formalism. For completeness, we provide here a quick overview of the mathematical formalism of posets and its application to the partial Borda count. Parts of what follows can be used as a review handout for students.

The partial Borda count generalizes to partially-ordered sets the ordinary Borda count for linear orders. Therefore, to give a meaningful treatment of the voting theory, the following mathematical topics should be introduced to the class in conjunction with the material on social choice theory.

A **poset** (short for **partially-ordered set**) is a set P together with a relation $<$ satisfying the following axioms:

- Irreflexivity: For all $x \in P$, $x \not< x$.

- Transitivity: For all $x, y, z \in P$, if $x < y$ and $y < z$, then $x < z$.

We write $x \leq y$ if $x < y$ or $x = y$. In the literature it is more common to see posets axiomatized in terms of the relation $\leq$, but for this module it is more natural to use $<$.

A relation $x < y$ in P is called a **cover relation**, and y is said to **cover** x, if there is no $z \in P$ such that $x < z < y$. Two elements of a poset are said to be **comparable** if one is less than the other. Otherwise they are **incomparable**. A poset in which any two elements are comparable is called a **chain**. A poset in which no two elements are comparable is called an **antichain**.

The standard way to visualize a poset is through its **Hasse diagram**, in which elements of the poset are drawn as nodes, and a line is drawn from one node x up to another node y whenever y covers x. Many examples of Hasse diagrams appear throughout Section 4 in the development of the module.

In the context of voting, a poset can be used to express a voter's prefences among a set A of alternatives. Thus for two alternatives $x, y \in A$, the statement "y is preferred over x" would be encoded by the relation $x < y$. The poset of preferences for an individual voter constitutes that voter's **ballot**, and the collection of ballots for all voters in an election is referred to as a **profile**.

The partial Borda count, as defined in [5] is based on assigning weights to alternatives in each ballot, and then aggregating the weights across all ballots and declaring the alternatives with the maximal weight to be the social choice. Weights are assigned to alternatives on individual ballots as follows. For a voter v whose ballot consists of some partial ordering $<_v$ of the alternatives A, the weight given to an alternative $a \in A$ on this particular ballot is

$$w(a) = 2 \cdot d(a) + i(a), \tag{11.1}$$

where $d(a)$ is the number of alternatives $b \in A$ such that $b <_v a$, and $i(a)$ is the number of alternatives that are incomparable with a. See Figure 11.2 for several ballots with weights calculated for each alternative. This method of assigning weights to alternatives and determining the social choice by aggregating weights is called the **partial Borda count**.

To understand equation (11.1), suppose there are n alternatives and each is given $n - 1$ points. Now require every alternative to give one of its points away to each alternative that is preferred over it. The number of points that an alternative $a \in A$ ends up with is precisely $2 \cdot d(a) + i(a)$. As an example to illustrate all of the ideas surrounding posets and the partial Borda count, we present the following profile with scores attached.

A.2 Overview of Social Choice Theory. Prior to the start of the class where students will work on the module, we provide them with a quick recap of the main ideas from Social Choice Theory. That recap is presented below as a handout for students.

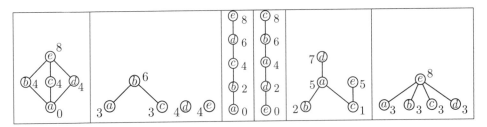

Figure 11.2. An Example of a Profile with Partial Borda Weights

Mathematics and Politics
Overview of Social Choice Theory

Voting with Two Alternatives. The main result in this area that we covered is **May's Theorem** and its generalization. In particular, if you want a social welfare function that is simultaneously

- anonymous (it is impossible to tell who voted for whom);

- neutral (each alternative has an equal chance of winning before voting begins); and

- monotone (if alternative **A** wins, and someone changes her vote from **B** to **A**, then **A** still wins),

then you must use a **quota system**. That is, an alternative wins precisely when they get a certain percentage (quota) of the votes – between 50 and 100 percent – otherwise the election is declared a tie. When the quota just exceeds 50, we call this **majority rules**.

Voting with Three or More Alternatives. The situation is much more complicated when there are three or more alternatives and **Arrow's Theorem** shows the extent to which our inherent notions of fairness are challenged. The main examples of social choice procedures that we learned are the following.

The Condorcet Procedure. An alternative wins if and only if they beat everyone else in a one-on-one election; otherwise there is no winner. This is an extremely strong requirement and is hard to meet.

Plurality. The most first-place votes wins. This ignores all rankings of the alternatives in the ballots below the top choice.

Anti-Plurality. The fewest last-place votes wins. This ignores all rankings of the alternatives in the ballots above the last choice.

The Borda Count. Given n alternatives, assign the first-place of each ballot $n-1$ points, the second place $n-2$ points, etc., until the last place on each ballot gets 0 points. The alternative with the most points wins.

The Hare System. Remove the alternative(s) with the fewest first-place votes and rerun the election; repeat the procedure until there is only one alternative left or more than one tied. These are the social choices.

The Coombs System. Similar to anti-plurality, this voting procedure deletes the alternative(s) with the most last-place votes and then reruns the election. Repeat this procedure until there is one alternative left or more than one tied. These are the social choices.

Sequential Pairwise Voting with a Fixed Agenda. Choose an ordering of the alternatives. Then run an election (majority rules) between the first two. The winner moves on against the third, and so on. The winner is the alternative who is left at the end.

Dictatorship. Choose a ballot at random. The top choice on that ballot is the winner.

Fairness. In class we studied five basic fairness criteria: **Always a Winner**, **Pareto**, **Monotonicity**, **Independence of Irrelevant Alternatives**, and the **Condorcet Winner Criterion**. The voting procedures above all satisfy some of these criteria (the specifics are worked out in your text, or in the homework). **Arrow's Theorem** says that the only voting procedure that satisfies Pareto, IIA, and Monotonicity is a dictatorship.

Other Examples.

Approval Voting. Each voter declares which alternatives they approve of. The most approvals wins.

The Partial Borda Count. Each voter submits a poset diagram of their true preferences (with ties and indecisiveness if applicable). The diagrams are then scored according to the following rule. If there are n alternatives, then each alternative is initially awarded $n - 1$ points. Then each alternative must give one point to every other alternative that is ranked higher. An example with five alternatives is the following:

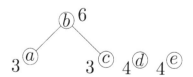

The alternative with the most points wins.

Lewis Carroll's Example. The alternative with the fewest swaps of adjacent alternatives needed to become a Condorcet winner is declared the winner. The problem with this method is its practicality: the computational complexity of deciding a winner gets at some of the hardest problems in theoretical computer science.

Ostrogowski's Paradox. This example illustrates the paradoxical nature of voting for individual issues versus the party line that can easily occur in local elections, even when there are only two parties.

There are two parties: **D** and **R**; five voters: **1, 2, 3, 4**, and **5**; and three issues: **Environment, Marriage Equality, Local Taxation**. Suppose each voter's true preferences are given in the following table:

Voter	Favorite on **Environment**	Favorite on **Marriage Equality**	Favorite on **Local Taxation**
1	D	R	R
2	R	D	R
3	R	R	D
4	D	D	D
5	D	D	D

If each voter votes for the party that best represents their views, then **R** wins, 3 to 2. On the other hand, **D** is the most-preferred party on all three issues.

Bibliography

[1] M. Ackerman, S. Choi, P. Coughlin, E. Gottlieb, J. Wood, "Elections with partially-ordered preferences", *Public Choice*, **157**, no. 1-2, 145-168 (2013)

[2] K. Arrow, "A Difficulty in the Concept of Social Welfare", *The Journal of Political Economy*, **58** (4), 328-346 (1950)

[3] Jean-Pierre Barthélémy, *Arrow's theorem: unusual domains and extended codomains*, Math. Social Sci. **3** (1982), no. 1, 79–89, DOI 10.1016/0165-4896(82)90008-7. MR665572

[4] Pavel Yu. Chebotarev and Elena Shamis, *Characterizations of scoring methods for preference aggregation*, Ann. Oper. Res. **80** (1998), 299–332, DOI 10.1023/A:1018928301345. MR1663399

[5] John Cullinan, Samuel K. Hsiao, and David Polett, *A Borda count for partially ordered ballots*, Soc. Choice Welf. **42** (2014), no. 4, 913–926, DOI 10.1007/s00355-013-0751-1. MR3196998

[6] Ronald Fagin, Ravi Kumar, Mohammad Mahdian, D. Sivakumar, and Erik Vee, *Comparing partial rankings*, SIAM J. Discrete Math. **20** (2006), no. 3, 628–648, DOI 10.1137/05063088X. MR2272220

[7] R. Niemi, "The Problem of Strategic Behavior under Approval Voting", *American Political Science Review*, **78**, 952-958 (1984)

[8] Alan D. Taylor and Allison M. Pacelli, *Mathematics and politics*, 2nd ed., Springer, New York, 2008. Strategy, voting, power and proof. MR2450599

[9] Richard P. Stanley, *Enumerative combinatorics. Vol. 1*, Cambridge Studies in Advanced Mathematics, vol. 49, Cambridge University Press, Cambridge, 1997. With a foreword by Gian-Carlo Rota; Corrected reprint of the 1986 original. MR1442260

[10] H. P. Young, *An axiomatization of Borda's rule*, J. Econom. Theory **9** (1974), no. 1, 43–52, DOI 10.1016/0022-0531(74)90073-8. MR0496492

[11] Special Section – 2013 American Mathematical Society Elections, *Notices Amer. Math. Soc.*, **60** (2013), no. 9 1071 - 1073.

12

Implementing Social Security: A Historical Role-Playing Game

John Curran and
Andrew Ross

Abstract. We introduce the *Ways & Means: 1935* role-playing game, which we use at Eastern Michigan University as part of our quantitative reasoning courses. The module explores distributive justice, including the rationale for and implementation of social insurance programs. In the game, students (as members of Congress) put together or argue against a Social Security law, re-imagining the historical debate held in the United States in 1935. In particular, students must determine what measures should be taken concerning old-age pensions, unemployment insurance, disability insurance, aid for the poor, and public health care, all of which were under serious consideration at the time. They must decide what is fair: who should receive payment, how much should be paid, and who should pay for the programs? Working in teams, students are expected to use historical data sets and quantitative arguments to develop workable social insurance schemes, and to support their points of view in the debate. (Students are encouraged to use a spreadsheet program to organize their analysis.) Typically, individual students are required to give a short speech and write a paper in support of their historical role's viewpoint. A final vote among the students determines the success and contents of the bill. To play the game, students use computation and numerical evidence to analyze what is fair and what is

feasible for social insurance programs. They also learn how to use numerical evidence in support of an argument, and to examine numerical claims
carefully. **Keywords:** Citizenship, descriptive statistics, distributive justice,
mathematical models, oral presentations, persuasive writing, role-playing,
social insurance.

The role-playing game presented here has students re-enact the historical debate
surrounding the passage of the Social Security law in 1935. It is appropriate for introductory courses in quantitative reasoning or statistics and takes two or three class
periods to complete.

Readers should note that the module described in this article is not self-contained
due to spatial constraints. Interested instructors may currently obtain all of the materials that support the game (free of charge) by emailing John Curran (`jcurran3@emich`
`.edu`) or Andrew Ross (`aross15@emich.edu`). The materials include two game books,
one for students and one for the instructor, including historical discussion, directions
for playing the game, role sheets, and worksheets. We also provide a spreadsheet containing historical data. Our colleagues Mark Higbee and Russell D. Jones (both in the
department of History and Philosophy at our university) are co-authors of the materials.

This module encourages students to formulate and analyze the quality of estimates
and mathematical models that are needed in implementing public policy. To do well
in this game, students must communicate their quantitative analysis clearly and per-
suasively. The camaraderie from being on a team encourages students to contribute as
best they can, and previously passive students may step forward as leaders. Because of
the fluid nature of the debate, an instructor can thoroughly assess students' conceptual
understanding and ability to apply concepts in novel situations.

1 Mathematical Content

The *Ways & Means: 1935* game is used in the 100-level mathematics course required for
most non-mathematics majors at Eastern Michigan University. The target course for
the game is a quantitative reasoning course, but it could be used as-is in an introductory
statistics course. With a change in emphasis, we believe it could also be used in an
introductory economics or political science course.

To play the game effectively, students need basic preparation in a number of mathematical topics commonly covered in quantitative reasoning classes. They need to use
large numbers (millions and billions), percentages, and percentage points accurately.
Ideally, students should be able to analyze a data set using graphical and numerical
descriptive measures, including scatterplots, correlation, mean, median, and percentages. In addition, students should be able to make Fermi-style estimates in cases where
the historical data are incomplete or in order to check the plausibility of an opposing
claim.

The game gives students experience using these basic mathematical skills in context. Students find that their level of skill will have a direct effect on how their team
performs in the game. If students are familiar with linear and exponential models, as
we require in our course, they will be able to apply these in a variety of ways, for example in the contexts of proportional and progressive taxation, tax burden, inflation,

interest rates, population growth, and economic growth. If students have been introduced to probability, as they are in our course, they can apply the ideas of expected value and variability to compare private savings plans with government-sponsored insurance. The game certainly works if students are not familiar with some of the latter material. Fundamentally, students must ascertain socioeconomic needs quantitatively and develop budgets or other plans to address them, and this can be done with fairly simple tools (as is often the case in practice!). If an instructor wishes, more advanced ideas in statistics, probability, and social choice theory can be introduced before the game to make the debate more sophisticated, or they can be introduced after the game in response to questions that arose during the game.

2 Context / Background

In addition to the quantitative tools mentioned in the previous section, students and instructors need to know the basic history of the Great Depression. This includes the economic situation, the political mood in the country, and the policies already in place by March 1935 (this is very important in keeping the debate focused and is addressed more below).

In 1935, the term "social security" referred to a broad range of possible social programs that might guarantee a basic standard of living for all Americans. The debate considers five social policies that did not exist at a national level in 1935:

Policy question 1. *Should the elderly receive guaranteed pensions and should workers' participation in the program be mandatory?*

This turned out to be the most important part of the Social Security Act of 1935, to the extent Americans now typically refer to public retirement benefits as "social security".

Policy question 2. *Should the unemployed be entitled to payments when they cannot find work?*

This was implemented, although with a number of exclusions that affected minorities and women disproportionately.

Policy question 3. *Should families be guaranteed a certain minimum income?*

In the end, aid was provided for dependent children but not for all families or individuals.

Policy question 4. *Should disabled individuals receive payments due to their disability?*

This did not make it into the final Social Security Act.

Policy question 5. *Are all Americans entitled to basic health care?*

The historical answer was effectively no, although there were some token provisions in the Social Security Act to study the problem further. The policy outcomes of running the module in class do not have to match the actual historical outcomes listed above.

The questions that come up during this module, including the ones we posed above, are relevant to a well-informed citizenry. Through this role-playing activity, students think critically about the types of social safety nets a society could and should offer its members, the various tensions related to such safety nets, and the compromises a society ends up making in order to implement significant policy changes.

Students are expected to represent a historical point of view in character, so the more they know about the period, the easier it will be to shape their arguments and analysis. For purposes of this game, reading the historical article given in the student game book and perhaps viewing some old film clips can provide sufficient background. The instructor may wish to give a brief introduction to the period, emphasizing the differences with modern points of view. The instructor manual provides sufficient historical material for this purpose.

This module grew out of conversations with history faculty at our university who are involved with the *Reacting to the Past* (RTTP) consortium. The consortium promotes the use of role-playing games in history courses. Members of the consortium obtained an NSF grant to extend the pedagogy to scientific fields. The grant led to development and play-testing of games for use in biology, astronomy, chemistry, environmental science, and health and nutrition courses. *Ways & Means: 1935* is the first game devoted primarily for quantitative reasoning and mathematics.

Commercially developed role-playing, board, and video games typically require several years of play-testing and development before they are complete. RTTP games are no different, and there is an active community of scholars engaged in developing such games. The consortium website (`http://reacting.barnard.edu`) is an excellent place to learn more about this pedagogy. *Ways & Means: 1935* has been play-tested at national conferences and used in classrooms during the last five years.

We have used *Ways & Means: 1935* over a dozen times in the classroom, in sections consisting of 20 to 30 students. Our university, EMU, is a large state institution with about 18,000 undergraduate students. The game has also been used by Professor Andrew Miller at Belmont University, a private liberal-arts college of 6,600 students. The principal restriction to using the game is section size. The number of students needs to be small enough so that everyone has a chance to speak and so that the teams can exchange ideas and develop a coherent plan. The module would be unwieldy in a course with more than 40 students. A small section is less of an obstacle to using the module, but the section needs to be big enough to allow for teamwork and negotiation. A class size of less than 10 is probably infeasible.

3 Instructor Preparation

Most of the preparation that an instructor needs is spelled out in the instructor manual for the game, which is described further below. It will take a few hours to read the entire manual, although some sections can be skimmed at first. As we mentioned above, instructors need to develop some comfort with the relevant history concerning the Great Depression. In this regard dates and names are less important than having a feel for what is appropriate language and behavior for members of Congress in 1935. It is especially important to be clear about which policies are on the table and which are not.

The policies considered in the game have been simplified to focus attention on the main issues, and the instructor needs a clear idea of what these should be.[1] By insisting that students must stick to a historical mindset, it is easier to keep them on topic — their options become clearer. For example, one of the times we used the game,

---

[1]An extensive discussion of the policies can be found in the students' and instructor's game books. A basic list of five policy questions was given in the previous section.

a number of students wanted to impose restrictions on the receipt of unemployment benefits (and aid for the poor) by requiring recipients to work for the state. Turning unemployment insurance into a jobs program was a significant distraction to classroom discussion, opening up all sorts of extraneous questions about how to implement such a program. It is possible to address such distractions in the role of an instructor or an in-game character (say, as a newspaper reporter, an observer from the congressional budget office, or outraged citizen), but the following observations helped get the class back on track:

- By 1935, several programs had been established by both the Hoover and Roosevelt administrations to provide publicly-funded jobs, covering a wide-variety of projects. Jobs programs were a done deal politically,[2] but social security was not.

- Insisting that those receiving benefits must work is more representative of the current political climate than the historical one, when despair about the economy and the absence of a safety net were keenly felt.

- Most importantly, talking about jobs programs misses the key question for distributive justice: *do those who cannot find work deserve benefits guaranteed by society, and if so, to what extent?*

As a first assignment, we usually ask students to read some background on the Great Depression and the New Deal, and then give a quiz about the reading. Possible readings include the Social Security Administration's web page (http://www.ssa.gov/history/), or for a more in-depth discussion, the historical appendix in the game book. A sample quiz on the historical context is included in the game book and in the appendix to this article. One of us (Ross) uses the results of this quiz to help assign roles, which is the next important step in preparing for the game.

The class is split into three factions: Liberals/New Dealers (usually Democrats), Conservatives/Anti-New-Dealers (usually Republicans), and the Center faction. The Liberals and Conservatives each try to convince the Center to vote for their proposals. Assigning roles is a novel task for most instructors. It is important to maintain numerical balance between the factions, which is straightforward, provided class enrollment is fixed. The factions should be of equal size, with any remainder assigned to the Conservatives, who have the more challenging goals. Their goals are more challenging because they are essentially trying to prevent any new social programs from starting, at a time when a lot of people want to take action to make things better. They are warning of future dangers of the proposed new programs, rather than solving present (1935) problems.

The harder question is about whether to assign leadership roles to stronger students, and our approaches to this question differ. Often each faction will choose a leader, and there are two special roles that must be assigned: Chairman of the Ways & Means committee and Eleanor Roosevelt. In particular, the Chairman has a lot of influence on the proceedings and a weak Chairman can be problematic. Curran prefers to assign roles at random (by drawing names out of a 1930s-style pill-box hat) and to coach the Chairman as needed prior to the first game session. Drawing names out of a

---

[2]It is true that the Supreme Court struck down a number of these programs, but their opposition can also be used to good effect. Instructors can inform the Congress that the Supreme Court will strike down any jobs program they pass, so they cannot receive any credit for it in the game.

hat is seen as fair and starts the process of getting students in a playful mood. Ross uses scores on a historical quiz to sort the students by apparent ability, and makes sure each faction has a fair distribution of students by ability. High-scoring students are selected for the special roles. This is explained to the class and is considered fair by the students.

The final decisions instructors must make before the game starts are the form of assessment to use and the consequences of winning. We recommend a mandatory speech (five minutes) and a short paper amplifying on the arguments given in the speech. Some instructors who use the RTTP method count winning and losing as part of the course grade; we prefer to base a student's grade on the speech and paper, and award extra credit for the performance of a student's team in the game. The possibility of winning or losing is highly motivating regardless of the actual stakes.

The role sheets that come with the instructor's game book provide a point system to determine whether a student has won or lost. Usually it is straightforward for an instructor to determine this, but there is pedagogic value in asking a student to explain in writing why they believe they have won. We use fairly simple rubrics to grade the speech and paper, categorized by effort, historical perspective, persuasiveness, and the use of quantitative evidence. Of course, instructors can easily use their own learning objectives and criteria.

Finally, we list a set of resources for the instructor. [1] offers a contemporary argument for implementing a variety of social insurance programs. The foreword to the book was written by Frances Perkins, then Secretary of Labor. [2] provides insight into the instructor experience using the RTTP method, and the design principles underlying the games. The Reacting to the Past website [3] is the place to start for instructors who are interested in the method, including introductory material, videos of students playing games, pedagogical discussion, and information on dozens of games at various stages of development. Free registration is required to access instructor materials. The Social Security Administration History website [4] provides an accessible introduction to the history of the Social Security Act. [5] gives an academic history of the Social Security Act.

4 The Module

The full materials needed for this module will not fit within the space available, but we provide the tables of contents in this section (see Figure 12.1) as well as some sample pages in the appendix. We are happy to send all of the materials for evaluation to interested instructors who email us (jcurran3@emich.edu or aross15@emich.edu). To save on copying, we would recommend that instructors post the material that students need on a password-protected course website.

We usually devote one class period to preparation, and one or two class periods to playing the game. In the preparation class, we introduce the historical context and the rules of the game, assign roles to the students, and provide time for the teams to meet, plan, and divide up tasks. The game book includes some worksheets on quantitative topics that we integrate into our courses earlier in the semester. However, if students seem particularly unprepared, the worksheets can be used for a second class period of preparation to give students a better idea of what they need to do while the teams continue to plan outside of class.

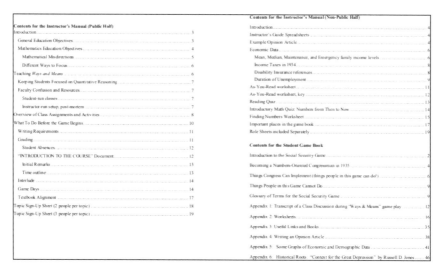

Figure 12.1. Contents of the game books.

Teams can proceed as they see fit, but it is often natural for them to assign 2 or 3 students to each policy, sometimes subdividing further by having some students prepare the team's proposal and others prepare to counter the other team. We typically require teams to post initial proposals to the course website before the debate begins. The students playing the chairman and the center faction can help in keeping the other teams on track here.

It takes roughly two hours of time to conduct the hearings on all five issues, preferably over two class sessions. This can be scaled back by reducing the number of issues that will be debated. If the discussion is limited to two issues, it can take place in less than an hour. Which topics to choose depends on the instructor's goals. In terms of social justice, the topics of disability insurance and aid for widows and orphans are similar in terms of overall rationale, and in terms of the quantitative work needed to describe the problem and establish a fair amount of payment. Health care and unemployment insurance can run along similar lines to the first two, but there are additional concerns in terms of whether those who use the programs or society at large should pay for them, and how much the programs should be modeled on insurance pools versus direct payments. Old-age pensions introduce yet another dimension of inter-generational distribution of resources. Historically, old-age pensions and unemployment insurance were the major components of the Social Security Act, and considering only those two would still lead to a rich debate. On the other hand, choosing to focus on only two of the other issues might simplify the level of analysis required.

Although not essential, there is definite value in spreading the debate over two class periods so that students can think about their opponents' arguments and prepare closing statements. The post-mortem can happen at the end of the hearings, or for short class periods, at the start of the following class period. During the post-mortem, it is customary in the RTTP community to go around the room and ask each student what they thought of the exercise, and for the instructor to point out the differences between what happened in the game and the historical outcome. (The game materials include a Power Point presentation instructors can use for this purpose.) It is not necessary to declare a winner at this point. Deferring the announcement allows for passions to subside so that students can consider the result objectively. Further, as noted in the previous section, it is worthwhile to ask each student to articulate in writing to the instructor why they believe they have won to encourage reflection.

Some students may find the prospect of doing a role-playing game in class daunting. We note that playing the game is not so much challenging as it is simply different from what students have experienced in mathematics classes. We like to emphasize that students have their teammates and a game book to help them out, and that playing the game can be fun. Supplying students with a sample opinion article (included in the instructor's guide) or making a sample speech can also help. Despite trepidation at the start of the game, students do figure out how to play fairly quickly. The act of playing is familiar from childhood, and players naturally smooth over a variety of potential complications. They key thing for an instructor is to ensure that the students direct the proceedings.

In role-playing jargon, the instructor using *Ways & Means: 1935* in their classroom is the *gamemaster*. A good gamemaster gives the players the impression that they are entirely in charge of what happens. The gamemaster only intervenes to smooth over unanticipated situations, which do arise since role-playing is inherently improvisational. It should be clear to everyone that the instructor's decisions concerning rules and scores are final, that the instructor can change the rules whenever they want to if the game is not developing in an appropriate direction, and that they can play the role of a minor character if one is called for. (Indeed, many students may consider the previous definition that of *instructor* rather than *gamemaster*.) We think it is best to adjust things with a light touch, so that changes go mostly unnoticed. If students think the game is rigged they will stop trying. Instructors can contact a student outside of class, or talk to the Chairman during a break, or circulate a "newspaper article" by email that contains hints between sessions. Instructors who are in touch with a former student who has played the game can bring them in as a coach for the factions and can pass information discreetly through this student to the participants.

All this can be a lot for an instructor to contemplate, but while a game is running it is usually clear how to proceed. In summary, our main pieces of advice for the instructor are

(1) **Read both halves (public and non-public) of the instructor manual**, focusing in particular on the role sheets. The roles are the source of most student questions.

(2) **Step back and let the students play.** They need to take their roles seriously, and so they need to have complete control over the proceedings. It is usually best not to interrupt the game even if a student makes a mathematical mistake that his opponents do not notice. Indeed, before the game starts we stress to the students that it is their job to catch their opponents' mistakes. Anything students miss can always be pointed out during the post-mortem.

(3) **Record the proceedings.** The discussion can become quite involved, and students can forget what they have already agreed upon. We have found the instructor is the best person to take notes and serve as a neutral reference — they can pretend to be a reporter or stenographer at the meeting. This reinforces the instructor's position as a neutral observer, and it facilitates assessing the outcome of the game.

(4) **Save time for a post-mortem.** There is lot to be learned from student comments, and the post-mortem can help students ramp down from being in character. The post-mortem is a teaching opportunity, when an instructor can point out things students did well and where they had difficulty.

5 Additional Thoughts

Our experience shows that college-level students are capable of playing this game to good effect. After all, everyone knows how to play! We note that the students taking EMU's quantitative reasoning course are below the national average in quantitative skills but are quite up to the game. There are students who flourish in this kind of activity when compared to traditional homework and exams, and students with stronger test scores who turn out to be weaker when applying the ideas in a more open context.

Some instructors have expressed concern that their colleagues may look askance at them for using a game to teach. We have found that our colleagues take our use of a role-playing game in the classroom seriously. Indeed, outside of the mathematics classroom, the use of role-playing, simulations, case-studies, and games to teach at the collegiate level is not uncommon. Further, administrators tend to view experimentation with teaching methods positively, and may actively support alternative methods, as has been the case at EMU.

The hardest aspect of designing the game was giving all of the factions a reasonable chance of winning. Historically, the conservatives did not have a chance, so we have strengthened them somewhat. The conservatives still have the harder task, but they do win in some classes. The ability of students within a faction has more impact than which faction they play.

We value the game especially for its ability to assess outcomes that can be hard to assess using exercises and exams. For example, the game allows us to evaluate our students' ability to apply quantitative ideas in context, and their ability to make a coherent argument using quantitative evidence.

Game-playing is inherently flexible, and instructors should be able to adapt the module easily to suit their own goals. As we mentioned above, the game can be scaled back if time is limited or the number of students is small. We have found that it can take some time for students to get used to playing a game, however, so scaling back too much may undercut the exercise. In fact, we have on some occasions run the same game twice in the same class, once very early in the semester and once towards the end. Students were very comfortable the second time around, and the debate and the outcome of the game was different from the first case. It might be possible to split the debate into five pieces distributed over the course of a semester, one for each issue. This would allow the mathematical prerequisites to be built up gradually over the course of the term.

There is clearly scope to use games like *Ways & Means: 1935* to address other social justice issues. For example, we have considered doing a similar game based on the Great Society legislation in the 1960s, in particular the expansion of welfare and the establishment of Medicaid. Social data collection was in its infancy in 1935, and on certain issues in our game students must make estimates due to the lack of hard data. The data available from thirty years later are more extensive, and might lead to a game with greater emphasis on data analysis than *Ways & Means: 1935*.

To this point, the game has been used primarily in our own classrooms. We encourage anyone interested in running the game to get in touch with us, and to share any ideas they may have. With a larger community of instructors using the game, it should be possible to perform some interesting pedagogical assessments and to develop new games based on that experience.

Appendix A: Sample Game Materials

A.1 Worksheet to Accompany the Historical Reading in the Student Game Book. Answers are provided in the instructor guide.

As-You-Read worksheet

(1) The NRA "President's Re-employment Agreement" called for minimum wages to be between $____ and $____ per week.

(2) The CCC paid $____ per month.

(3) Huey Long called for an income guarantee of $____ per year per household.

(4) Francis Townsend called for a pension of $____ per month for every American over age 60.

(5) Now fill in this table comparing how much the plans would pay each person in the plan. Use these facts:

   • there are 48 work hours per week (most people worked 6 days),

   • there are 52 weeks per year,

   • there are 4.33 weeks per month.

Also, for the NRA, use the average of the lower and upper values that you answered in #1.

Plan	Per Hour	Per Week	Per Month	Per Year
NRA				
CCC				
Long				
Townsend				

Keep in mind, though, that they are talking about different portions of the population. Men employed by the CCC were typically single and young, while the NRA and Long plans were more for the general working class, and the Townsend plan was for seniors.

(6) For the Townsend plan, calculate how much it would cost the country each year, assuming that people over age 60 were 8.5% of the population of 127 million people.

A.2 Historical Quiz. Answers are provided in the instructor guide.

Reading Quiz

(1) Social Security was founded during which span of time? Circle the correct answer.

    (a) Right after the Civil War

    (b) Just before World War I

    (c) Between the two World Wars

    (d) During World War II

    (e) After World War II

(2) Social Security was founded during which span of time? Circle the correct answer.

    (a) Half a dozen years before the big stock market crash

    (b) A few months before the big stock market crash

    (c) A few months after the big stock market crash

    (d) Half a dozen years after the big stock market crash

(3) Social Security was founded during a time called what? Circle the correct answer.

    (a) The Roaring Twenties

    (b) The Long War

    (c) The Cold War

    (d) The Long Winter

    (e) The Great Depression

(4) Who was the president when Social Security was founded (first and last name)?

    _____

(5) Was he in the Democratic or Republican party? (circle one)

(6) What was the name of the First Lady (the president's wife)?

    _____

(7) Was the First Lady:

    (a) substantially less active in politics than most first ladies up to her time,

    (b) about as active in politics, or

    (c) substantially more active?

(8) Social Security was part of a big batch of programs that was called _____

(9) The unemployment rate when Social Security was founded was about:

    (a) 1%

    (b) 5%

    (c) 20%

    (d) 50%

    (e) 75%

The following questions are not answered in the game book; they require only a little bit of online searching, and are not essential to playing the game.

(10) Are women allowed to vote at the time our game is set? Circle one:

Allowed   Not Allowed

(11) Is alcohol still prohibited in the U.S. at the time our game is set, or has Prohibition been repealed? Circle one:

Still prohibited   No longer prohibited

(12) Which of the following technologies existed back then? Circle Yes or No as appropriate:

(a) Yes / No : electric power in cities

(b) Yes / No : telephones

(c) Yes / No : automobiles

(d) Yes / No : airlines (for rich people)

(e) Yes / No : airlines (for many people, like today)

(f) Yes / No : movies

(g) Yes / No : radio

(h) Yes / No : television

A.3 Sample Pension Worksheet. Answers are provided in the instructor guide.

1935 Pension Idea, Version 2: Still a bit simplified

Many proposals for the benefit level do not use a simple percentage of the income. Instead, there are various different percentages that apply, depending on which income bracket you are in (the same concept, but not the same numbers, as income-tax brackets). Here is a proposed rule for determining the monthly benefit ("Primary Insurance Amount" or PIA):

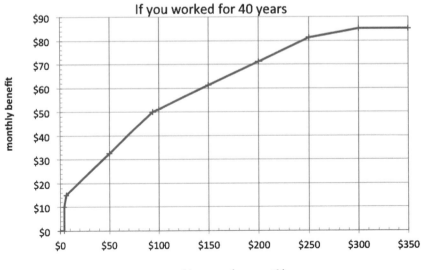

(1) Based on the graph above, please fill out the blank spots in this table, for a highly simplified sample of 10 households (income amounts chosen to roughly match the whole population). You should make sure you understand where the first PIA amount ($35.83) comes from, so you are on the right track. You will read some values from the graph; do not be too concerned about precision.

Yearly Income	Monthly Income	Households	PIA	Yearly benefit per household	Yearly total	PIA as % of monthly income
$700	$58	6	$35.83			
$1730	$144	3				
$5285	$440	1				

Total : $5763

(2) If the total yearly payout is roughly $5763 for these 10 households, what would the yearly total payout be for 3.3 million retired households?

Here is a graph of PIA/monthly income before retiring. The numbers in your last column above should match this graph:

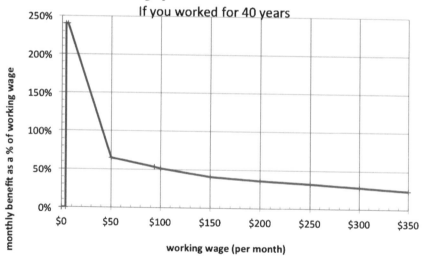

(3) Now compute the total paid in by each household (at a Social Security tax rate of 4.34% on the first $3000/year, and 0% on the amount over $3000, for forty years) and the total paid out in benefits (for ten years). The number of households of each type is not important.

Yearly Income	Total tax paid in 40 years	PIA	Yearly benefit per household	Total benefits paid over 10 years	Total benefits/ Total tax paid
$700		$35.83			
$1730					
$5285					

Notice that each type of household is getting back well more than they put in, if they survive for ten years in retirement. How does this work in the long run?

A.4 Student Role Sheet for Eleanor Roosevelt. ELEANOR ROOSEVELT

You are a leader of the Liberal New-Dealer faction.

You are the First Lady of the United States, Eleanor Roosevelt (born 1884). Your husband Franklin has been president since March 1933, and already you are, by far, the most influential female to ever live in the White House. In the future, scholars will argue that you, as First Lady, held more political power than any other woman in American history prior to the late 20th century.

You are one of your husband's most trusted and influential advisors, and for much of the 1920s, your income and public profile were larger than Franklin's. Yet your marriage is complex — on levels personal and political all at once. Without a doubt, you are a mover and a shaker, a radical reformer who never steps out in front of her husband. The President often counts on you for advice and for information on numerous issues that your own broad contacts can gather more quickly than his official channels. But like many husbands, he doesn't always follow his wife's advice.

You never criticize the President's choices or decisions publicly, yet you lean on him privately and seek to persuade him to make your views his own. You favor a more expansive social policy for the New Deal than he does. FDR respects and admires your political talent and instincts. If he is to move beyond the social security proposals already made by the Administration, it will only be, most likely, if you convince him that more expansive policies are both capable of being passed by Congress and better remedies for national problems. The President is not going to come out for a plan that isn't politically viable, and you wouldn't want him to, either. But you do want a more systematic bill to be passed and signed into law than what is currently before the congressional committees.

As a liberal New Dealer, you want Congress to pass a Social Security law that does all of the following:

- Aid for poor families, especially widows and orphans.

- Old age pensions — and for all working people.

- Health insurance for all American families.

- Disability benefits for all workers.

- Unemployment insurance.

You are less concerned about how to pay for these benefits than are others. Still, as a practical political matter, you and your allies need to credibly propose a means of paying for these benefits. FDR and most party leaders are very worried about federal deficits, as both an economic problem and as a political-electoral liability. Running for president in 1932, FDR promised to balance the budget, and he hasn't yet.

For twenty years, you have been part of a circle of social reformers and friends who've advocated a universal social security system — one that would cover everybody, and provide protections to the most needy of Americans; anything less than a truly universal system of security will be a defeat.

While you would never say anything that makes your husband look bad, you can offer proposals that he has not. In particular, you care most about universal health insurance.

You are not a member of Congress so you cannot vote directly. *Your influence is strong enough, however, that you do control one vote via a trusted congressman. For purposes of the game, you will vote with the rest of the participants.*

VICTORY POINTS*

As the First Lady, to win the game you must obtain a personal victory. Still, what is good for the liberal New-Dealer team is good for you, so work with them. Without their help, you cannot hope to win.

As the wife of the President, your reputation is unimpeachable, and you will be unaffected by other players' penalty points. Also, you will start the game with a political favor card that you can use to persuade a wavering centrist or open-minded conservative to vote in the best interests of the nation, as you see them. You may earn additional favor cards if you earn bonus points.

If the final legislation includes …

Old Age Pensions

Defined-benefit pensions for all workers, paid for by taxes: 150 points
Defined-contribution pensions for all workers, paid for by taxes: 75 points
Pensions, but with restrictions on the workers eligible: 100 points if defined-benefit, 25 if defined-contribution

Unemployment Benefits

Benefits at 150% of subsistence level for all unemployed workers, paid for by taxes: 100 points
Benefits at subsistence level for all unemployed workers, paid for by taxes: 50 points
Unemployment insurance, but with restrictions on the workers eligible: 75 points if benefits at 150% of subsistence level, 25 points if at subsistence level

Disability Benefits

Lifetime income support at full salary for those disabled while working, paid for by taxes: 75 points
Lifetime income support at half salary for those disabled while working, paid for by taxes: 50 points

Aid for the Poor

Enough income to live on for all poor families, paid for by taxes: 150 points
Enough income to live on for widows and orphans only, paid for by taxes: 50 points

Health Care

Free health care available to all Americans, paid for by taxes: 300 points
Mandatory health insurance for all Americans, paid for by individuals according to a sliding scale: 200 points
Voluntary health care plan with some government involvement: 75 points

Investing Tax Dollars in the Stock Market

Are you nuts? Minus 200 points.

Bonus Points*:

A strong speech, including quantitative analysis: 50 points
Good questions/answers during Q&A: 20 points
Showing leadership: 40 points
Using your own historical research: 60 points

Note: you can earn points multiple times in these categories.

Penalty Points*

Breaking historical character: minus 5 points
Showing your role sheet to anyone else: minus 10 points
Not understanding your faction's goals: minus 10 points
Rude or disruptive behavior: minus 20 points

Note: you can earn penalty points multiple times.

VICTORY OBJECTIVE: You must earn 600 points.

*The Gamemaster may assign bonus and penalty points at his/her discretion. The Game-
master may adjust the points awarded as s/he sees fit.

Bibliography

[1] Abraham Epstein, *Insecurity, a Challenge to America: A Study of Social Insurance in the United States and Abroad*, Smith & Haas, New York, 1933.

[2] Nicolas Proctor, *Reacting to the Past Game Designer's Handbook*, 3rd ed., CreateSpace Independent Publishing Platform, 2011.

[3] Reacting to the Past website, https://reacting.barnard.edu/, accessed on August 21, 2016.

[4] Social Security Administration History website, http://www.socialsecurity.gov/history/, accessed on August 21, 2016.

[5] Edwin E. Witte, *The Development of the Social Security Act*, University of Wisconsin Press, Madison WI, 1963.

13

Matching Kids to Schools: The School Choice Problem

**Julie Glass and
Gizem Karaali**

Abstract. The problem of assigning students to schools in public school districts is complicated both mathematically and socially. One must balance the needs, preferences, and priorities of schools, cities, and families, while providing kids with the best possible education. Choosing a process by which school assignments are made introduces questions about social justice via the ideas of access, entitlement, and advocacy. School districts often use complex mathematical matching mechanisms in an attempt to address these questions while providing the best possible opportunities for the greatest number of students. Throughout, fairness is used as a proxy for questions of social justice. This lesson introduces the *School Assignment Problem* which seeks "fair" solutions to this challenging problem. Students learn about a variety of well-established matching algorithms and use novel approaches to applying these algorithms in the setting of school assignments. In the process they learn some basics of game theory and mechanism design. The accompanying problems ask that students think deeply about fairness in designing and applying these algorithms and, at the same time, be encouraged to ask why it is that a safe and academically rigorous public school environment has become a scarce resource in so many settings. **Keywords:** Education, equity, access, fair division, game theory, mechanism design, cooperative learning.

This module introduces students to the foundational definitions, concepts, ideas, and applications of game theory and mechanism design in a social justice context. Specifically, we use the *School Choice Problem* as an example and vehicle for exploring the properties of matching mechanisms and notions of fairness. Notions of fairness are used throughout as a proxy for questions of social justice including access, entitlement, and advocacy.

The School Choice Problem is concerned with matching students to schools in a given district and can be thought of as an example of managing scarce resources with equity as a primary driver. Straightforward examples are used to illustrate key properties of several commonly studied (and used) mechanisms to show how each attempts to balance competing desirable outcomes. Examples given show relationships between and among some of the most commonly sought properties. These properties include (Pareto) *efficiency*, *stability*, and *fairness*, applied in a rigorous and relevant setting. Concrete and meaningful applications of abstract notions will strengthen students' understanding of the value of seemingly esoteric constructs. It is of note that these same tools and criteria are often used to analyze human behavior and can be thought of as a way to study rational choice. Thus other applications in the allocation of scarce resources with social justice in mind will be clear.

1 Mathematical Content

This module is appropriate for use in courses in discrete or finite mathematics, courses that explore the "nature of mathematics", or an introduction to proof/transition to advanced mathematics courses, depending on the extent to which background material is presented and explored. It can be a part of a game theory unit or a unit on social choice theory. Alternatively it can stand alone as a concrete context for proofs or example-building exercises. See [10] for a textbook introducing the idea of mathematical proof through the use of the stable marriage problem.

There are minimal prerequisites in terms of mathematical content as most ideas and notation are introduced within the module. Suggestions will be made as to how much depth is appropriate for the investigations in different courses.

In particular, we use set-theoretic notation in most of our definitions, so if students are not familiar or comfortable with it, instructors might consider alternative ways to introduce the necessary constructions. Typically, concrete examples (see Section 4.5) will make the notions sufficiently clear even if precise mathematical definitions may not be appropriate.

2 Context / Background

The 80s and 90s saw the rise of school choice programs that engaged parents and caregivers in identifying and requesting more desirable schools for their students. Historically the use of neighborhood schools defined by walk zones was thought to encourage and promote school attendance and greater community commitment to public education. However, a quick examination reveals that school performance often reflects economic and social disparities; this observation has caused districts to reconsider school assignment processes. It is also possible that more competition for students from charter schools and the voucher movement (purporting to make independent schools within reach for previously excluded families) motivated districts to make changes to their school assignment processes.

The main objective of any school choice system is to allow students to attend schools that are more desirable (for whatever reason) than the school for which they are zoned. The sad state of affairs is that for many districts across the country there are more students than available spots in desirable schools. From a social justice perspective the best way to "solve" the school choice problem would be to develop an environment where there were no "bad" schools so that all students would have access to the education they deserve. However, absent the (mathematical) capacity to develop this solution, we seek instead to find ways to make the most desirable matchings efficiently and effectively.

Economists have been studying the school choice problem almost since school choice was introduced; see for instance [2,3,5,6,11]. Instead of looking at school choice as a policy problem, economists have used mechanism design to develop matchings given the seemingly intractable problem of too few seats in desirable schools. The main idea of mechanism design is to consider "metagames" with a game-designer. Games are "designed" with an eye towards creating a situation in which the players disclose their private information and also which results in a desired (by the designer) outcome. Game theory and mechanism design attempt to describe and model human behavior using mathematical tools. Questions of social justice provide natural areas of investigation and application.

Though we did not use the module in its current form in our own classes, we think it can be used at any type of institution. A class size of no more than 35 students would be ideal.

3 Instructor Preparation

All relevant notation and basic terms used are given in Section 3.1. Mechanisms used in the module are introduced in Sections 3.2–3.5. Once the instructor is familiar with this material, it can be presented to students in class or provided in a handout.

Using the tools and language introduced below, the School Choice Problem seeks to develop the most advantageous matching of students to schools. The district is viewed as a designer while the students (or those who represent them) are viewed as players. The goal is to create a school choice mechanism that will allocate the available resources (seats in schools) among the players (students) subject to district priorities and legal requirements.

For the instructor, a good starting point is the paper [2]. One of the earliest works addressing the school choice problem mathematically, this paper also provides sufficient background to explore the problem. A rich source of opinions backed with some solid data is Diane Ravitch's blog https://dianeravitch.net/; see for instance her post on October 30, 2015, available at http://dianeravitch.net/2015/10/30/nsba-research-shows-that-school-choice-does-not-improve-education/.

Alternatively instructors may see [12] for some arguments for and against school choice.

3.1 Notation and Basic Terms Used. Each definition below can be introduced in several ways depending on the level of the students and instructor preference.

Definition 1. *Let I denote a non-empty set of students, and S a non-empty set of schools. A **matching** M : I → I × S is a function that associates every student with exactly one school. We will also use the related function M : I → S and write M[i] = s if M(i) = (i, s).*

Instructor Insight: The function defined above may be somewhat difficult to understand, and students could certainly spend some time thinking about how it behaves. In particular, this definition can be used to reinforce standard notions around functions, including notation, domain, range, and other properties such as one-to-one and onto.

Definition 2. *For all s ∈ S, let q_s denote the (maximum) **capacity** of s. Then we will have: $|\{i \in I : M(i) = (i, s)\}| \leq q_s$. When the inequality is strict, we say that s is **unfilled**.*

Instructor Insight: Students may observe that this definition implies that it makes sense for some schools to go unfilled in order to keep other schools sufficiently filled. It might be a good idea to open this up a bit and ask them followup questions such as, "Is it 'fair' for one school to be full while another does not have enough students to be viable?" and "Would this perpetuate existing inequalities?" Students might be asked then to consider whether there might be a place in the process to ensure some "evening out" of enrollment.

Definition 3. *Define the **ranking function** ϕ_i : S → ℕ of a student i ∈ I by letting $\phi_i(s)$ denote i's ranking of s ∈ S. In other words $\phi_i(s) = j$ if s is i's jth choice school.*

If i prefers s_k to s_l, we write $s_k \succ_i s_l$, or simply $s_k \succ s_l$ if i is unambiguous. Note that throughout this paper the notation $\succ$ denotes a strict order; if we want to describe a weak order, we will write $\succeq$. When all schools appear in a student's preference profile, we say that the profile is **complete**, otherwise we say it is **incomplete**. We denote a set consisting of preference profiles for each student in I by $\mathbf{P} = \{P_i : i \in I\}$.

For students familiar with some set theory, the notion of a partition may be introduced along with the following more general definition.

Definition 4. *A **preference profile** for a student i ∈ I, written P_i, is a tuple $(S_1, ..., S_n)$ where the S_j's form a partition of S and every element of S_j is preferred to every element of S_k if and only if j < k.*

Instructor Insight: Prior to introducing this definition formally, the previous paragraph could be used to motivate the notion of a partition. If students have already studied partitions, see if they identify this as an example of one. If students have not studied partitions, see what they come up with.

Students can also be asked about the efficacy/appropriateness of a strict ordering.

From here forward, each time we refer to a school ranked jth by student i we could instead refer to the set of such schools, S_j, and modify the explanation accordingly.

Definition 5. *A **priority structure** for a school s ∈ S, written Π_s, is a tuple $(i_1, ..., i_n)$ where the i_j's are an ordered list of all students in I and a student i_j is preferred over student*

i_k if and only if $j < k$. If school s prefers i_k to i_l we write $i_k >_s i_l$, or simply $i_k > i_l$ if s is unambiguous, where $>$ denotes a strict order and $\geq$ a weak order. We denote a set consisting of priority structures for each school in S by $\Pi = \{\Pi_s : s \in S\}$.

Definition 6. A matching M' (**Pareto**) **dominates** M if $M'[i] \succeq_i M[i]$ for all i and $M'[j] \succ_j M[j]$ is strict for some j. Informally, this means that in a matching that Pareto dominates, all students are at least as well off and at least one student is strictly better off than in the dominated matching. That is, nobody is worse off and at least one person is better off.

Definition 7. A (**Pareto**) **efficient** (or (**Pareto**) **optimal**) **matching** is a matching that is not (Pareto) dominated. Again, informally, this means that there is no other matching in which someone has a better outcome unless someone else ends up worse off.

Definition 8. Let Π_s be a priority structure for school s. We say that a matching M **violates the priority** of $i \in I$ for s if there exist some $j \in I$ and $s' \in S$ such that

(1) $M[j] = s$, $M[i] = s'$, i.e., student j gets assigned to school s under M and student i gets assigned to school s' under M,

(2) $s \succ_i s'$, i.e., student i would rather attend school s than school s', and

(3) $i \succ_s j$, i.e., school s would rather have student i than student j enroll as a student.

Definition 9. We say that a matching M is **stable** if

(1) M does not violate any priorities.

(2) No student is matched to a lower-ranked school when a strictly preferred school is unfilled.

Instructor Insight: This implies that a desirable school will be filled while a less desirable school may be severely underfilled. Students can discuss whether or not this is "fair" as mentioned above. Should all schools be guaranteed some critical level of enrollment?

3.2 The Stable Marriage Problem and the Gale-Shapley Deferred Acceptance (DA) Mechanism.
Given n men and n women, the goal of the classical *Stable Marriage Problem* (SMP) is to pair men and women together so that there are no two people in different groups who would both rather have each other than their current partners.[1] Such a matching is said to be *stable*. In 1962, David Gale and Lloyd Shapley proved that, for any equal number of men and women, it is always possible to solve the SMP and make all marriages stable [9]. The solution is accomplished via the mechanism described below.

---

[1] We recognize the anachronistic nature of this outdated notion of marriage. It turns out that the optimal algorithm in its original context is also quite heteronormative and stereotypical about the roles of men and women in a heterosexual relationship. Both issues might make interesting discussion topics for the classroom if an instructor wishes to bring up the roles of historical context and bias in mathematical exposition. Nonetheless, the mathematics is interesting in its own right, and the methods discussed to address the problem will apply to any two-sided matching problem; see for instance medical internship match-ups [14] or kidney transplant markets [15].

The Gale-Shapley *Deferred Acceptance* (DA) algorithm proceeds in rounds as follows:

Round 1: Each man proposes to his most preferred woman and the woman says "maybe" to her highest ranked proposer and "no" to any others. The accepted pairs are now engaged.

And in general:

Round k, k ≥ 1: Each unengaged man proposes to his next choice partner (to whom he has not yet proposed) and the women respond as in Round 1, accepting their most preferred suitor (including any partner to whom she is already engaged) with a "maybe" and rejecting the others (which might include the guy to whom she had already been engaged).

The algorithm will terminate with a matching that is stable.

3.3 The Student Optimal Stable Matching (SOSM) Mechanism.

The Gale-Shapley Deferred Acceptance algorithm was adapted to the School Choice Problem in the form of the *Student Optimal Stable Matching* (SOSM) mechanism. SOSM is widely held to be a practical mechanism for implementation. In particular, New York City and Boston [3–6] have adopted SOSM as their mechanism of choice. SOSM runs as follows:

Round 1: Each student applies first to their first choice school. Each school then tentatively accepts the student(s) highest on its preference list among those who applied that round (such students are now waitlisted) and rejects the rest beyond its quota. We remove each waitlisted student from the market. All unwaitlisted students move on to the next round.

And in general:

Round k, k ≥ 1: Each unassigned student applies to their next choice school. Each school considers the new applicants together with the current waitlist and repopulates the waitlist with those applicants who are highest on its priority list and rejects the rest beyond its quota. We remove each waitlisted student from the market. All unwaitlisted students move on to the next round. The algorithm runs until all students have been assigned.

3.4 Top Trading Cycles (TTC) Mechanism.

Next we describe the *Top Trading Cycles* (TTC) mechanism which was originally introduced in [16] (also see [1]) and adapted to the school choice context in [2] as an alternative to the SOSM. TTC is a strategy-proof mechanism which compromises on stability to achieve efficiency. The mechanism proceeds as follows:

Round 1: Each student points to their first choice school. Similarly each school points to its first choice applicant. Since there are finitely many students and schools, there is at least one cycle. For each such cycle do the following: Assign each student in the cycle to the school they are pointing to and remove the student and the school from the market. All unassigned students and unfilled schools move on to the next round.

And in general:

Round k, k ≥ 1: Each unassigned student points to their top choice school among the unfilled ones. Each unfilled school points to the student whom it ranks highest among the unassigned students. There should be at least one cycle. For each such cycle do the following: Assign each student in the cycle to the school they are pointing to and remove the student and the school from the market. All unassigned students and unfilled schools move on to the next round. The algorithm runs until all students have been assigned.

Thus the essence of the TTC in the school choice context is that once in a trading cycle, students are allowed to trade placements among themselves.

3.5 Efficiency Adjusted Deferred Acceptance Mechanism (EADAM).

This section includes more advanced material, and may be skipped if instructors prefer.

The examples presented in Section 4.5 illustrate that SOSM's strict adherence to stability can result in a loss of efficiency. That is, it is possible to Pareto dominate the SOSM outcome thus improving the overall well-being of students while violating one or more of the school priorities. A mechanism called the Efficiency Adjusted Deferred Acceptance Mechanism (EADAM) was proposed in [11] as a way to alleviate some of the efficiency costs of stability by iteratively running SOSM and modifying the preferences of any interrupters (who displace another student only to be ultimately displaced themselves) such that the SOSM outcome is Pareto dominated. As any Pareto domination of SOSM will lead to priority violations (cf. [9]), EADAM leads to at least one priority violation.

In order to understand EADAM we must first define an **interrupter**. Let student i be one who is tentatively placed in a school s at some step t while running the SOSM, and rejected from that same school at some later step t'. If there exists at least one other student who is rejected from school s after step $t - 1$ and before step t', then we call student i an **interrupter** for school s and the pair (i, s) is an **interrupting pair** of step t'. An interrupter is **consenting** if they allow the mechanism to violate their priorities at no expense to them. The EADAM then runs as follows:

Round 0: Run the SOSM.

Round 1: Find the last step (of the SOSM run in Round 0) at which a consenting interrupter is rejected from the school for which they are an interrupter. Identify all interrupting pairs in that step which contain a consenting interrupter. If there are no such pairs, then stop. Otherwise, for each identified interrupting pair (i, s), remove school s from the preference list of student i without changing the relative order of the remaining schools. Rerun the SOSM with the new preference profile for i until all students have been assigned.

And in general:

Round k, k ≥ 1: Find the last step (of the SOSM run in the previous round) at which a consenting interrupter is rejected from the school for which they are an interrupter. Identify all interrupting pairs in that step which contain a consenting interrupter. If there are no such pairs, then stop. Otherwise for each identified interrupting pair (i, s), remove school s from the preference list of student i without changing the relative order of the remaining schools. Rerun the SOSM with the new preference profile until all students have been assigned.

Even though the end result of consenting for interrupters is that they allow the mechanism to violate their priorities for schools they would not be assigned to anyway, the step-by-step description above points toward a different route to obtaining the same outcome. That is, the practical outcome would be the same if the consenting interrupters were to modify their preference lists in such a way as to drop schools that they'd not have been assigned to anyway. Thus, instead of asking students to sign consent forms to waive priorities, as would be required to run the EADAM, we could in theory ask them to reconsider their preference lists.

4 The Module

In this module we use the School Choice Problem to explore properties of matching mechanisms and notions of fairness. The complete module takes three-to-seven class meetings, with some preliminary and followup work to be done out of class.

This section is organized according to the parts of the module, as follows:

Section 4.1: describes an activity that students should complete prior to the mathematical portion of the module. This can be assigned as homework about one week in advance of the first class meeting at which the topic will be discussed. Instructors may announce this as preparation for an upcoming class discussion or assign a more formal write-up associated to it.

Section 4.2: introduces the School Choice Problem through a series of questions for discussion. We suggest these to be used in the first in-class session for the module.

Section 4.3: introduces the notation, basic terms, and definitions required to understand the common mechanisms described later. Instructors may choose to spend some class time for this content, or, alternatively, they can use a handout to introduce the material, to be reviewed briefly in a follow-up class meeting.

Section 4.4: introduces the common mechanisms used in the School Choice Problem. Again, instructors may choose to spend some class time for this content, or alternatively, depending on time constraints, they can use a handout to introduce the material, to be reviewed briefly in a follow-up class meeting. We suggest using at least one class hour on each mechanism. If time is limited, distributing the three mechanisms to student groups so that each group focuses on one mechanism will help.

Section 4.5: provides a series of examples that illustrate some of the strengths, weaknesses, and tradeoffs of the mechanisms as applied to the School Choice Problem. Instructors can use these as templates for the specific examples they choose to present in their classes whenever necessary.

4.1 The Set Up. Students should be asked to do background research on material related to the social justice issue under consideration. In particular, before the first mathematical session, students should explore:

(1) **The school assignment *priorities* for their home town (or local) district.** If multiple districts are investigated, a wiki or online discussion board can be used to collect the frequency of each priority. Commonly used priorities are walk zone, sibling attending the school, special programming, and various combinations of these. Districts may use different vocabulary to present their school assignment process. Students should consider looking for phrases such as "assignment", "lottery", "options", and "choice". Generally a phone call to the district will provide a webpage or other documentation that describes the overall process from the point of view of the family and will include information about the district priority structure. Students might also be asked to simply guess and/or propose priorities for schools. If a student's home district has only one school per level, they can be asked to do this assignment for a nearby district that does have some choice. Or, they could investigate the process in a community outside the United States.

It is also interesting to ask students why they think these (actual or proposed) priorities are relevant to a fair distribution of school seats. What is the community trying to accomplish with the various priorities, what values do they reveal and/or represent? Many districts will have information about their vision and goals and how their assignment process helps them achieve their core values in educating their students (see, for example, `http://www.berkeleyschools.net/information-on-berkeley-unifieds-student-assisignment-plan/`).

(2) **The socioeconomics of a neighborhood compared to the performance of the walk-zone school.** We simply ask our students to think about this, without requiring in-depth analysis, but instructors may choose to develop this segment further as a significant component of the preliminary assignment. In that case, we suggest the following websites as starting points: `http://www.census.gov` and `http://www.greatschools.net`.

(3) **Websites that model matching mechanisms in various social contexts.** Students may seek to find simple implementations of the stable marriage problem in well-known contexts, such as the annual medical internship match-ups or kidney transplant markets. For instance the following websites have different but interesting explorations of the Stable Marriage Problem:

`http://mathworld.wolfram.com/StableMarriageProblem.html`

`http://mathsite.math.berkeley.edu/smp/smp.html`

(4) **Political party and/or specific candidate messaging around education and school choice.** Political candidates for city offices inevitably talk about public school quality. Can any patterns be uncovered as to how these political statements translate into priority structures for school assignments?

(5) **Other situations where "matchings" reveal social injustice and whether or not there are ways for communities to "solve" these problems through social programs and community organizing.** Some possible examples are environmental injustice (e.g., poorer neighborhoods being more polluted), access to healthy food, and political district gerrymandering.

4.2 The School Choice Problem. To start the in-class portion of the module, the instructor should introduce the social justice context of (school) choice as described in Section 2. In addition, any information that was gathered via the wiki or other repository in Section 4.1 should be shared.

Below is a brief outline of how we would move the discussion forward, via the introduction of formal notions, each followed by a discussion question. If the course is entry-level, some elementary set theory and notation should be introduced to supplement the content here.

Introduce the Stable Marriage Problem (SMP): Given n men and n women, the goal of the SMP is to pair up men and women so that there are no two people in opposite groups who would both rather have each other than their assigned partners. Such a matching is said to be *stable*. Students might be asked to experiment with strategies for a given desirable outcome. They can start with just two in each group, and move to three. Small examples of efficient solutions will be demonstrated in Section 4.5 once vetted mechanisms have been presented.[2]

Discuss: *Is this a reasonable definition of "stable"? Does it necessarily mean that everyone is happy?*

At this stage, students may be encouraged to think of other ways to mathematically define a stable matching (or a happy marriage or a perfect match).

Introduce the School Choice Problem (SCP): Given a population of students who have ranked the schools in their district and a collection of public schools with set capacities and enrollment priorities decided by the district, match students with seats in schools so that "fairness" is maximized by some relevant measure.

As before, small examples might be useful here. See Section 4.5 for some example templates. For the first few examples in class, especially for more entry-level courses, we recommend making up specific names for students and schools.

Discuss: *What, if any, are the differences between the SMP and the SCP?*

Some students might point toward the seeming symmetry of the SMP (both sides are individuals looking for partners) as opposed to the asymmetry of the SCP (students and schools are not even the same kind of entity, and, in particular, schools are matched with multiple students). This point becomes important when the Gale-Shapley DA algorithm is introduced and then is adapted to the SCP as SOSM.

4.3 Definitions, Notation, Terminology. At this point, students should be introduced to the basic definitions, but not the mechanisms, provided in Section 3.1. These can be provided via a handout or presented in class. Each definition can be presented mathematically, followed by a discussion of how it aligns with more informal notions of fairness. More specifically students should be encouraged to rephrase in their own words, the mathematical definitions provided, and then discuss how these formal notions relate to the more informal notion of fairness.

---

[2]This problem is historically presented as a pairing between men and women. Instructors who wish to avoid the heteronormative implications might prefer to present the problem using any two disjoint sets of objects for whom a matching is desirable. For example, two sets of swing dancers with one set preferring to lead and the other preferring to follow, with no gender correspondence noted, would work just as well.

Next, it is time, using this formal language, to reintroduce the School Choice Problem. In the context of mechanism design, the district is viewed as a designer while the students (or those who represent them) are viewed as players. The goal is to create a school choice mechanism that will allocate the available resources (seats in schools) among the players (students) subject to district priorities and legal requirements.

An appropriate discussion point at this stage is whether a focus on priorities (i.e., stability as represented by the district designer) as opposed to other criteria is "fair". That is, if enforcing district rules leads to undesirable choices for students, what then? Students should explore how, if at all, the concepts defined by the formal language align with their ideas about fairness.

4.4 Mechanisms. Finally each of the mechanisms from Section 3.1 should be introduced. This can be handled in a couple of different ways, depending on the amount of time available and also the audience. One possibility is to take two weeks (six class hours total) to introduce, discuss, and practice with each mechanism (two class hours per mechanism). Another possibility is to break up the class and give each group one mechanism to study, understand, and then present to the rest of the class (thus two class hours to work and a third to present, for all of the mechanisms). A quick outline for how to approach each mechanism is as follows:

a) Definition.

b) Worked examples constructed from templates in Section 4.5.

c) Examples developed by class.

d) Social justice implications and relationship to fairness.

We strongly recommend making up names for the students and schools in the first few examples so the class can follow the steps of the algorithms described more easily.

Below are relevant questions to ask for each of the presented mechanisms. Ultimately, much of the interesting discussion arises via comparison of the mechanisms. Each set of questions includes the mathematical as well as social justice implications.

Questions related to the Gale Shapley Deferred Acceptance (DA) mechanism and the Stable Marriage Problem.

(1) *Is this notion of stable the same as the one defined in Section 3.1?*

(2) *How do we know this process is finite? Why will it end with a stable matching?*

(3) *Is the result fair to all students? What priorities did you find or can you think of that would result in an improved outcome from a social justice perspective?*

Questions related to the Student Optimal Stable Matching (SOSM) mechanism.

(1) *Does this mechanism seem fair?*

(2) *Can you see any way the system might be manipulated by districts and/or families?*

(3) *Does everyone have an equal shot at their desired school?*

Questions related to the Top Trading Cycles (TTC) mechanism.

(1) *Are the requirements for participation in the TTC reasonable?*

(2) *Might some populations fare poorly because of insufficient access to information about schools or the process itself?*

Questions related to the Efficiency Adjusted Deferred Acceptance mechanism (EADAM). (Advanced material)

(1) *What are the tradeoffs between stability and efficiency?*

(2) *Is it reasonable to favor stability over efficiency or vice versa?*

(3) *Are there different winners and losers in each case?*

4.5 Examples, Examples, Examples. In this section we provide template examples of specific school choice problems and their solutions via one of the mechanisms introduced above. We examine the characteristics of the outcomes as measured by common criteria. Note that the priority and preference profiles are given, as well as the outcomes of the relevant mechanisms. We suggest using specific names for students and schools, at least at the beginning, in particular if the students are not comfortable with mathematical notation.

The corresponding exercises can be made more challenging if the instructor chooses to ask students to calculate the resulting matching themselves. In addition, for each example, students could be asked to come up with (different) sets of priority and preference profiles that illustrate the stated characteristics of the outcome.

Example 1: Matchings under SOSM can be both stable and Pareto efficient.
Assume there are two individual students, i_1 and i_2; two schools, s_1 and s_2; and one spot at each school. The preferences and priorities are given by:

$$s_1 : i_1 \succ i_2, \quad i_1 : s_1 \succ s_2,$$
$$s_2 : i_2 \succ i_1, \quad i_2 : s_2 \succ s_1.$$

Under SOSM, the matching is:

$$\begin{pmatrix} i_1 & i_2 \\ s_1 & s_2 \end{pmatrix}$$

which is both Pareto-efficient and stable.

Followup Question: Prove that / explain why the outcome above is, indeed, the outcome of SOSM and that it is Pareto-efficient and stable.

Example 2: SOSM can also produce a stable matching that is not Pareto efficient. Assume there are three schools, s_1, s_2, s_3, and three individual students, i_1, i_2, i_3. The priorities of the schools and the preferences of the students are given by:

$$s_1 : i_1 \succ i_3 \succ i_2, \quad i_1 : s_2 \succ s_1 \succ s_3,$$
$$s_2 : i_2 \succ i_1 \succ i_3, \quad i_2 : s_1 \succ s_2 \succ s_3,$$
$$s_3 : i_2 \succ i_1 \succ i_3, \quad i_3 : s_1 \succ s_2 \succ s_3.$$

Here, SOSM finds the only stable matching:

$$\begin{pmatrix} i_1 & i_2 & i_3 \\ s_1 & s_2 & s_3 \end{pmatrix}.$$

However, this matching is Pareto-dominated by:

$$\begin{pmatrix} i_1 & i_2 & i_3 \\ s_2 & s_1 & s_3 \end{pmatrix}$$

which is not stable because student i_3 has *justified envy* of student i_2. In other words, student i_3 prefers school s_1 to school s_3 and school s_1 prefers student i_3 to student i_2.

Followup Question: Prove that / explain why the first matching above is, indeed, the outcome of SOSM and that it is stable but not Pareto-efficient.

In fact, SOSM Pareto-dominates all stable matchings, and thus any Pareto domination of an SOSM matching must not be stable (cf. [**9**]).

Example 3: A given SCP may have multiple stable matchings. Assume there are three schools, s_1, s_2, s_3, and three individual students, i_1, i_2, i_3. The priorities of the schools and the preferences of the students are given by:

$$s_1 : i_2 \succ i_1 \succ i_3, \qquad i_1 : s_2 \succ s_3 \succ s_1,$$
$$s_2 : i_1 \succ i_2 \succ i_3, \qquad i_2 : s_3 \succ s_2 \succ s_1,$$
$$s_3 : i_3 \succ i_1 \succ i_2, \qquad i_3 : s_2 \succ s_1 \succ s_3.$$

Here, there are two stable matchings:

Stable Matching 1:   Stable Matching 2:

$$\begin{pmatrix} i_1 & i_2 & i_3 \\ s_2 & s_1 & s_3 \end{pmatrix}. \qquad\qquad \begin{pmatrix} i_1 & i_2 & i_3 \\ s_2 & s_3 & s_1 \end{pmatrix}.$$

Stable Matching 2 Pareto dominates Stable Matching 1.

Followup Question 1: Show that both outcomes above are stable and that Matching 2 Pareto dominates Matching 1. Which one is the outcome of SOSM?

Followup Question 2: Which matching do you think is better? Why?

Example 4: The TTC outcome may Pareto-dominate the SOSM matching. Assume there are three schools, s_1, s_2, s_3; three individual students, i_1, i_2, i_3; and only one seat at each school. The priorities of the schools and the preferences of the students are given by:

$$s_1 : i_1 \succ i_3 \succ i_2, \qquad i_1 : s_2 \succ s_1 \succ s_3,$$
$$s_2 : i_2 \succ i_1 \succ i_3, \qquad i_2 : s_1 \succ s_2 \succ s_3,$$
$$s_3 : i_2 \succ i_1 \succ i_3, \qquad i_3 : s_1 \succ s_2 \succ s_3.$$

SOSM finds this matching:

$$\begin{pmatrix} i_1 & i_2 & i_3 \\ s_1 & s_2 & s_3 \end{pmatrix}$$

and TTC finds a matching that Pareto dominates the SOSM matching:

$$\begin{pmatrix} i_1 & i_2 & i_3 \\ s_2 & s_1 & s_3 \end{pmatrix}.$$

Followup Question 1: Prove that the outcomes are as stated above.

Followup Question 2: Which matching do you think is better, why?

Advanced Followup Question: Find the EADAM outcome of the SCP above.

Note that (i_3, s_1) is an interrupting pair and the EADAM with the consent of i_3 outputs the Pareto efficient matching produced by TTC.

Example 5: The TTC outcome does not always Pareto-dominate the SOSM matching. Let $I = \{i_1, i_2, i_3, i_4\}$ (the set of individual students) and $S = \{s_1, s_2, s_3, s_4\}$ (the set of schools) where each school has only one seat. The priorities for the schools and the preferences of the students are given as follows:

$$i_1 : s_1 \succ s_3 \succ s_2 \succ s_4, \quad s_1 : i_4 \succ i_3 \succ i_1 \succ i_2,$$
$$i_2 : s_1 \succ s_2 \succ s_3 \succ s_4, \quad s_2 : i_2 \succ i_3 \succ i_1 \succ i_4,$$
$$i_3 : s_2 \succ s_1 \succ s_3 \succ s_4, \quad s_3 : i_2 \succ i_3 \succ i_1 \succ i_4,$$
$$i_4 : s_4 \succ s_3 \succ s_2 \succ s_1, \quad s_4 : i_1 \succ i_2 \succ i_3 \succ i_4.$$

The matching under SOSM is:

$$\begin{pmatrix} i_1 & i_2 & i_3 & i_4 \\ s_3 & s_2 & s_1 & s_4 \end{pmatrix}$$

while the matching under TTC does not Pareto-dominate it:

$$\begin{pmatrix} i_1 & i_2 & i_3 & i_4 \\ s_1 & s_2 & s_3 & s_4 \end{pmatrix}.$$

Followup Question 1: Prove that the outcomes are as stated above.

Followup Question 2: Which matching do you think is better? Why?

Example 6: SOSM and TTC can output different Pareto-efficient matchings. Assume there are three schools, s_1, s_2, s_3; three individual students, i_1, i_2, i_3; and there are two seats at s_2 and one seat each at s_1 and s_3. The priorities of the schools and the preferences of the students are given by:

$$s_1 : i_1 \succ i_3 \succ i_2, \quad i_1 : s_2 \succ s_1 \succ s_3,$$
$$s_2 : i_2 \succ i_1 \succ i_3, \quad i_2 : s_1 \succ s_2 \succ s_3,$$
$$s_3 : i_2 \succ i_1 \succ i_3, \quad i_3 : s_1 \succ s_2 \succ s_3.$$

SOSM finds a Pareto-efficient matching:

$$\begin{pmatrix} i_1 & i_2 & i_3 \\ s_2 & s_2 & s_1 \end{pmatrix}$$

and TTC finds a different Pareto-efficient matching:

$$\begin{pmatrix} i_1 & i_2 & i_3 \\ s_2 & s_1 & s_2 \end{pmatrix}.$$

Followup Question 1: Prove that the outcomes are as stated above.

Followup Question 2: Which matching do you think is better? Why?

5 Additional Thoughts

Notions of fairness are complex and difficult to analyze but they arise in many different contexts related to social justice. There are multiple pathways students might explore fairness mathematically. Below are just a few examples of pairing this module with other relevant content.

A political choice path: This module on school choice may be augmented by a module on social choice and voting schemes. See Chapters 11 and 16 in this volume for relevant content.

A fair division path: The n-person cake cutting problem may be another interesting complement to this module. See [7, 13, 17] for more on cake cutting and [8] for more on other examples and constructs related to fair division.

A districting path: Another mathematically rich districting problem is that of forming of congressional districts (gerrymandering). See Chapter 9 in this volume for relevant content.

Students interested specifically in the social justice issues related to the School Choice Problem may wish to follow up with

- investigating how charter schools or voucher programs impact enrollment at public schools;

- investigating how charter schools or voucher programs impact test scores at public schools;

- watching controversial documentaries such as *Waiting for Superman* and the rebuttal *The Inconvenient Truth Behind Waiting for Superman*.

Once again [12] provides rich background and context on the related debates.

Bibliography

[1] A. Abdulkadiroğlu and T. Sönmez. "House allocation with existing tenants". *Journal of Economic Theory*, Volume **88** (1999), pages 233–260.

[2] A. Abdulkadiroğlu and T. Sönmez. "School choice: A mechanism design approach". *The American Economic Review*, Volume **93** Issue 3 (2003), pages 729–747.

[3] Atila Abdulkadiroğlu, Parag A. Pathak, and Alvin E. Roth. "The New York City high school match". *American Economic Review - Papers and Proceedings*, pages 364–367, May 2005.

[4] Atila Abdulkadiroğlu, Parag A Pathak, and Alvin E Roth. "Strategy-proofness versus efficiency in matching with indifferences: Redesigning the NYC high school match". *American Economic Review*, Volume **99** Issue 5 (June 2009), pages 1954–1978.

[5] Atila Abdulkadiroğlu, Parag A. Pathak, Alvin E. Roth, and T. Sönmez. "The Boston public school match". *American Economic Review - Papers and Proceedings*, pages 368–371, May 2005.

[6] Atila Abdulkadiroğlu, Parag A. Pathak, Alvin E. Roth, and T. Sönmez. "Changing the Boston school-choice mechanism: Strategy-proofness as equal access". *Working paper*, May 2006.

[7] Steven J. Brams and Alan D. Taylor, *An envy-free cake division protocol*, Amer. Math. Monthly **102** (1995), no. 1, 9–18, DOI 10.2307/2974850. MR1321451

[8] Steven J. Brams and Alan D. Taylor, *Fair division*, Cambridge University Press, Cambridge, 1996. From cake-cutting to dispute resolution. MR1381896

[9] D. Gale and L. S. Shapley, *College Admissions and the Stability of Marriage*, Amer. Math. Monthly **69** (1962), no. 1, 9–15, DOI 10.2307/2312726. MR1531503

[10] Eric Gassett, *Discrete Mathematics with Proof*, 2nd edition, Wiley, Hoboken, NJ, 2009.

[11] Onur Kesten. "School choice with consent". *Quarterly Journal of Economics*, Volume **125** Issue 3 (2010), pages 1297–1348.

[12] Diane Ravitch, *The Death and Life of the Great American School System: How Testing and Choice Are Undermining Education.* Basic Books, 2011.

[13] J. M. Robertson and W. A. Webb. *Cake-Cutting Algorithms: Be Fair If You Can.* A K Peters Ltd., Natick, Mass., 1998.

[14] Alvin E. Roth and Elliott Peranson. "The Redesign of the Matching Market for American Physicians: Some Engineering Aspects of Economic Design". *The American Economic Review*, Volume **89** Issue 4 (September 1999), pages 748–780.

[15] Alvin E. Roth, Tayfun Sönmez, and M. Utku Ünver, *Pairwise kidney exchange*, J. Econom. Theory **125** (2005), no. 2, 151–188, DOI 10.1016/j.jet.2005.04.004. MR2186970

[16] Lloyd Shapley and Herbert Scarf, *On cores and indivisibility*, J. Math. Econom. **1** (1974), no. 1, 23–37, DOI 10.1016/0304-4068(74)90033-0. MR0416531

[17] Walter Stromquist, *How to cut a cake fairly*, Amer. Math. Monthly **87** (1980), no. 8, 640–644, DOI 10.2307/2320951. MR600922

14

Modeling the 2008 Subprime Mortgage Crisis in the United States

Bárbara González-Arévalo and
Wanwan Huang

Abstract. We study the 2008 subprime mortgage crisis in the United States. Affordable housing, home ownership, and predatory lending are all social justice themes that are explored during the course of the project. Students have an opportunity to investigate this crisis and what caused it. They work in groups, with each group choosing a different subpopulation to study (for example, a specific age group, a specific racial group, or a specific socioeconomic group). As part of the project, they find real data on the subpopulation of their choosing. They then use the financial mathematics knowledge learned in class to analyze the mortgages of their subgroup, and study different scenarios such as a short sale, foreclosure, etc. Students need to know how to perform interest rate and inflation calculations, present and future value calculations, mortgage amortization set-up, and calculations for various mortgage types. Towards the end of the semester, students prepare a poster for a university-wide poster session, and they also make an informational flyer to be distributed to their school/community/neighborhood.
Keywords: Financial mathematics, Calculus 2, mortgages, interest rates, group work, posters.

We describe a seven-part module that can be used in most courses in financial mathematics in twelve consecutive weeks. Each of the seven parts is due approximately every two weeks, so the students are engaged in the project throughout the semester. At our institution, the students are sophomores or juniors who have taken *Calculus 2* and an introductory-level finance course. If they have no finance background, then the instructor needs to introduce the basic finance terms in class. The project makes heavy use of the financial mathematics taught throughout the semester, and so it would not work in a *Calculus 2* class with students who do not have a financial background.

1 Mathematical Content

This module will work best for a *Financial Mathematics* course (also called *Theory of Interest* in some institutions). This is a course that has *Calculus 2* as a prerequisite, mainly because it uses series. Ideally, students should have taken at least *Calculus 2* and some introductory level finance course to prepare them for this project. They should understand the concepts and basic terminology before they start. Then the project could be run in two different ways:

(1) For more advanced students who are currently taking a financial mathematics course, the mathematical background would be assumed and they would be required to set up and perform all the pertinent calculations.

(2) For less advanced students who have taken *Calculus 2* and an introductory finance course, the instructor could provide Excel sheets pre-loaded with the necessary calculations.

In either case, although at different levels, the mathematical skills of figuring out loan repayment will be applied to finish the project. Such topics may include amortization method, retrospective and prospective formulas for outstanding balance, non-level interest rates, capitalization of interest, amortization with level payments, interest only with lump sum payment at the end, and sinking fund method [2]. In this module, students learn how to apply these concepts to a real-life situation, and thus learn about the implications of the different types of loan repayment methods.

2 Context / Background

The recession of 2008 in the United States was triggered by the subprime mortgage crisis [4]. Consumer spending was down, the housing market had plummeted, foreclosure numbers continued to rise, and the stock market had been shaken. The subprime crisis and resulting foreclosure fallout had caused dissension and social justice problems among consumers, lenders, and legislators and spawned furious debate over the causes and possible fixes of the "mess" [1].

This module was developed in 2009, in the midst of the crisis, for use in a liberal arts college, where class sizes do not exceed 30 students per section (usually with around 20 students). It was originally developed and taught with the support of an NSF/SENCER (National Science Foundation/Science Education for New Civic Engagement and Responsibilities) Implementation subaward in conjunction with Dr. Jie Yu.

The typical population in the *Financial Mathematics* class consisted mainly of actuarial science majors in their sophomore or junior years. A great proportion of these students were first-generation college students, coming from a wide variety of backgrounds. These were highly motivated students, but they tended to have a somewhat weak mathematical background. Most of them wanted to be professional actuaries, so this project would give them a solid experience to talk about during an internship or job interview. It would also help them put into practice all the concepts they were learning in class.

3 Instructor Preparation

The instructor should have a basic economic understanding of the subprime mortgage crisis; reading references [1] and [4], for instance, should be sufficient. The initial part of the project, described in more detail below, will involve the students researching and answering some background questions on the subprime mortgage crisis itself. The instructor could also show the movies *The Big Short* [6] or *Margin Call* [3]; alternatively students could watch them before class.

3.1 Before the first class meeting.
Based on the financial mathematics knowledge in a textbook such as [2], the instructor should complete the project herself/himself before giving it out in class. It is of particular importance to learn what data can actually be found from the sources (e.g., [5,7]), in order to better guide the students in their search (sometimes suggesting a different subpopulation or a different time period).

It is also essential for the instructor to become acquainted with the technology that will be needed for the project. Students may find the technology particularly challenging, so the instructor needs to feel very comfortable with it. Depending on the resources available at their institution, the instructor should set up at least one library session for their class. Ideally, during this session, the students would work with a librarian, learning how to search for data and background information needed for the project.

3.2 During class.
It is essential to give students some time during class to work on the project. This has two purposes: it conveys to the students the importance the instructor is giving to the project, and it forces them to start early and have group discussions, as opposed to just "dividing" the work and never actually working together for a cohesive project. Also, giving them some time to have a group discussion in class will help them to get to know each other and work together after class. During the discussion, the instructor should walk around and join in to answer specific questions or give suggestions.

As will be explained in more detail below, the groups themselves should vary throughout the semester. Among other things, this helps with otherwise insurmountable group dynamic problems. It also prevents a group from being dismantled by the withdrawal date, and it helps less outgoing members of the groups to have more meaningful interactions.

3.3 Post assignment: Grading issues.
Grading should occur at each stage of the project. No single assignment should be worth too many points, but as a whole, the project should be a significant part of the course grade. There should be some group grading as well as individual grading. The instructor can incorporate one or two questions related to the project in some or all of the exams. For the essays or posters, the

instructor could let the students do the grading for each other based on a standard grading policy. This will help the students stress the key learning points again and motivate them to get involved in the process.

Some suggestions for grading the different parts of the project follow:

Reflective Journals: These can be graded on a 0/1 scale: completed/not completed. The most important aspects of the reflective journals are for students to put their thoughts in writing and to ensure active participation of all students.

Parts 1, 2, 4, 5, and Final Paper: Instructors who feel uncomfortable grading essays could pair up with a composition instructor at their institution. Many institutions are now focusing on writing across the curriculum, so there could be many resources available to help faculty with this.

Part 3: The grading of this part should focus on checking that the students have collected the appropriate data they will need to be able to complete the rest of the project.

Poster and Flyer: The grading here should focus on how well they managed to present all the relevant information. Did the students do a good job disseminating their results?

4 The Module

This module requires using financial mathematics concepts to *model the 2008 U.S. subprime mortgage crisis*. It consists of seven parts and a set of reflective journal questions, all of which are given in full detail in Appendix A.

Assignment	Content	Due Date	Grade
Reflective Journals	Varied	Every week	10%
Part 1	Literature review on subprime mortgage crisis	Week 3	10%
Part 2	Mortgages and subpopulation to study	Week 5	15%
Part 3	Real data collection on the subpopulation	Week 7	10%
Part 4	Preliminary analysis of the data	Week 10	15%
Part 5	Deeper analysis of the data, conclusions	Week 12	15%
Poster and Flyer	Informational flyer, poster presentation	Week 13	10%
Final Paper	Complete project	Week 14	15%

Part 1: Individually, students research the subprime mortgage crisis and develop an understanding of what caused this crisis and how it started. Students are asked to go to the library and do a literature search. This search is guided by questions given to the students: *What is a mortgage? What was the 2008 U.S. subprime mortgage crisis? How was it different from the internet bubble crisis? When did the crisis start? What might have caused this crisis? Who have been affected by this crisis and how?*

Part 2: In groups of two, students research all the components of a mortgage and decide on a relevant subpopulation to study, for example:

- a specific age group (e.g., households where the mortgage holder is less than 30 years old),
- a specific racial group (e.g., Hispanic households),

- a specific socioeconomic group (e.g., households making between $40,000 to $60,000),

- etc.

Students should choose a subpopulation that means something to them. They need to find data on their specific subgroup, and should work closely with a librarian to do this. If there is no librarian available at their institution, the instructor will have to guide the students through this process.

Since different people in the class will choose different subgroups, at the end of the semester there will be an opportunity to discuss how the subprime mortgage crisis affected different subpopulations.

Part 3: In groups of three, students find real data on their chosen subpopulation. Possible databases for the students to use:

- *MedianMortgages(2003);*
- *MedianMortgages(2008);*
- *MedianMortgages(2009);*
- *HighEndMortgages;*
- *Median Household Income by State.*

Students can find the above databases or similar databases from sources like [**5**] or [**7**].

Part 4: In groups of three, students write an essay based on the data and their knowledge about mortgages. The essay needs to contain answers to the following questions: *How reliable do you think your data is and why? Can any real-life data be 100% reliable? What could lead to under-reporting or over-reporting of cases? What average/median annual income do members of your subpopulation make? Choose different values for the components of a mortgage and calculate each individual's monthly payment. How much total interest do they need to pay over the life of the mortgage? What percentage of each income is the mortgage payment? Is this amount reasonable? What can you conclude from these results?*

Part 5: In groups of three, students analyze the mortgages based on short sale and foreclosure scenarios.

Poster and Flyer: In groups of three, students prepare a poster of findings and make a flyer to share the knowledge they gained with their school/community/neighborhood.

Final Paper: In groups of three, students write a paper based on all the work they have done. This should involve bringing everything together into a final product of their work throughout the semester.

5 Additional Thoughts

5.1 On using this module for a different class. Even though one might be tempted to use this project in a finite mathematics class, we do not believe this would be appropriate. By design, this project involves complicated calculations (of the sort taught in financial mathematics at the level of the Society of Actuaries Exam FM or the Casualty Actuarial Society Exam 2). It might be possible to use it in a *Business Calculus* class if most of the students have a strong finance background, and the mathematical content is modified as suggested for less advanced students.

5.2 On making this project shorter. This project was designed to take all semester. The main benefit from this is that it simulates what students will need to do to complete a project in the workplace. Having deadlines throughout the semester helps the students complete a much higher quality paper at the end. Instructors might be worried that there is not enough time in the semester to include such a comprehensive project, but this is simply not our experience. In our opinion, it would be preferable to leave out a small amount of material than to not do the project. In practice, however, we have not actually needed to do this; we cover all the expected material. We follow the *Interest Theory* part of the syllabus from the Society of Actuaries for Exam FM, which can be found at `https://www.soa.org/education/exam-req/edu-exam-fm-detail.aspx`, accessed on September 2, 2016.

5.3 On student perceptions of the project. Students tend to think that the project is too much work, so it is important for the instructor to keep reminding students why they are doing the project. They are practicing the skills they will actually need in the workplace, which include group work, applying the mathematics they are learning to real-life situations, and disseminating their results in written and oral formats. It is also good to remind students this experience will be great for job interviews.

5.4 On how to form project groups and work with group dynamics. Group work is often dreaded equally by students and instructors. The main reason for this is that it creates conflict. After many years of requiring group work in our classrooms, we have settled on a few good practices to try to avoid the worst of these conflicts:

(1) Instructors should not settle on groups at the beginning of the semester. There are many reasons for this; but mainly, students do not know each other yet, the instructor does not know the students yet, some students will eventually drop the class, etc.

(2) As explained in Appendix A, we start by assigning an individual piece. This will guarantee that every student starts thinking about the project and does some research on their own. On the second part of the project we have them work in groups of two. This eases their reluctance to work in groups, and they know they do not need to keep working with this person if the dynamic does not work.

(3) The ideal group size is 3. We obviously allow groups of 4 or 2 if the number of students in the class is not a multiple of 3, but we force students to be in groups of 3 as much as possible (that is, only one group of 2 or 4 is necessary).

(4) After Part 3 (see Appendix A), we allow the students to rearrange the groups one last time. This should happen after the final withdrawal date. That way, the final groups of 3 actually have three people in them. And by this time almost everybody is working well with their chosen group.

(5) Instructors should make sure to inquire about group dynamics throughout the semester and intervene to solve any serious problems early on.

5.5 On the reason for the poster and the flyer. An important part of teaching students about social justice is civic engagement. Students need to become involved with their communities. Doing a project involving social justice that only an instructor is going to read becomes meaningless. Students need to take what they have learned about social justice, and they need to do something about it. Disseminating their knowledge is the very first step in this, and so the poster and the flyer are essential parts of the learning objectives. If an instructor is more ambitious, she could have her students send letters to their local government, to regulatory agencies, or to banks, explaining what they found out and suggesting an action plan to prevent another crisis.

Appendix A: Assignments and Handouts

These assignments can be distributed electronically (through a learning management system for instance) or can be printed out and distributed.

Part 1: To be done individually. You will study the subprime mortgage crisis. First of all, you need to have an understanding of what caused this crisis and how it started.

Using the library, math, economics, or finance professors, and/or reputable internet sources, research the following questions and write an essay. Don't simply answer the questions; write a good essay about this topic. You will use this essay in your final paper. Make sure to correctly cite all your references; plagiarism will not be accepted.

(1) *What is a mortgage?*

(2) *What was the 2008 U.S. subprime mortgage crisis? How was it different from the internet bubble crisis?*

(3) *When did the crisis start?*

(4) *What might have caused this crisis?*

    (a) *Were the homebuyers misguided?*

    (b) *Were the banks sufficiently regulated?*

    (c) *How about the leverage of the investment banks and hedge funds?*

    (d) *Other explanations?*

(5) *Who have been affected by this crisis and how?*

    (a) *Homeowners/homebuyers;*

    (b) *Banks/financial institutions;*

    (c) *Schools/universities;*

    (d) *Other sectors;*

    (e) *Other countries.*

Part 2: To be done in groups of 2. You will study all the components of a mortgage and decide on a relevant subpopulation to study. Each group may choose a different subpopulation (for example: a specific age group, a specific racial or ethnic group, a specific socioeconomic group, etc.). Remember to write everything in essay form and correctly cite all sources. This will all be part of your final project paper.

(1) *What are the components of a mortgage?*

(2) *What are the common values of these components? (constant/variable)*

(3) *How does the monthly payment change as some component changes?*

(4) *What is your subpopulation?*

    (a) *What do they have in common?*
    (b) *How are they different from others?*
    (c) *Will you be able to get information on this group?*
    (d) *Do they take on mortgages or not?*
    (e) *Why did you choose this group of people?*

Part 3: To be done in groups of 3. Now you need to find real data on the sub-population you chose. Different members of your group will probably have chosen different subpopulations in Part 2. This is good! Now you can choose one of them, making sure you can actually find the data you need. Put the data into an Excel file. Remember to cite your source and to keep track of what exactly the numbers represent. You will need all of this in subsequent parts of the project.

Part 4: To be done in the final groups of 3. Based on the data you obtained (you can use the data chosen by any of the members of your group) and the knowledge you have about mortgages, answer the following questions and write an essay. Don't simply answer the questions; write a good essay about this topic. You will use this essay in your final paper. Make sure to correctly cite all your references. Plagiarism will not be accepted.

(1) *How reliable do you think your data is and why? Can any real-life data be 100% reliable? What could lead to under-reporting or over-reporting of cases?*

(2) *Use your data to answer the following:*

    (a) *What average/median annual income do members of your subpopulation make?*
    (b) *Choose different values for the components of a mortgage, and calculate each individual's monthly payment.*
    (c) *How much total interest do they need to pay over the life of the mortgage?*
    (d) *What percentage of each income is the mortgage payment? Is this amount reasonable?*
    (e) *What can you conclude from these results?*

Part 5: To be done in the final groups of 3. Analyze mortgages based on two scenarios (short sale and foreclosure). Answer the following questions and write an essay. You will use this essay in your final paper. Make sure to correctly cite all your references.

(1) *What are the definitions of short sale and foreclosure?*

(2) *What are the components of a short sale position or foreclosure position?*

(3) *Under which condition will someone have a short sale or foreclose? And why?*

(4) *Which scenario does your data fit? And why?*

(5) *Who will benefit from a short sale or foreclosure?*

(6) *What kind of conclusion can you draw from the analysis?*

Poster and Flyer: To be done in the final groups of 3. Prepare a poster of your findings and also make a flyer to be distributed to your school/ community/neighborhood. It should be clear and understandable to people who haven't learned financial mathematics before. All of the data/graphics you borrowed should have their source cited.

For the poster, the text and graphics should be readable from at least three feet away. Your poster should look professional; use color judiciously and make sure spelling and grammar are correct. Personalize your poster. All graphics should be related to the topic, and all borrowed graphics should have their source cited. Your poster should include the following:

(1) A title, the name of everyone in your group, the name of the university, and the date.

(2) An introduction to the topic (both from the business perspective and mathematically).

(3) A statement about the data used (where did it come from?). Do not put the actual data in here.

(4) Your model with a description of all the terms.

(5) Appropriate graphs.

(6) A discussion of what you learned from your model and what you hope to teach the general public.

(7) Citations for all sources used (both for the data and for everything else!).

You will present your poster during a poster session at the end of the semester. You should be prepared to discuss how you came up with your model and answer any questions people may have. You will distribute the informational flyer in your school/community/neighborhood, so make sure it looks professional and is easy to read.

Final Paper: To be done in the final groups of 3. Create a paper based on all your work this semester. Your paper should have the following:

A title page: This should include the title, authors, and date.

An abstract: This is a brief summary of the paper. Usually it is one or two paragraphs long.

An introduction: This is a brief summary of the business and the mathematical backgrounds and rationale for the model.

Data collection: This describes what your data is and where it came from. Do not put the actual data in here.

Methodology: This is a summary of how you arrived at your final model.

Discussion of the model: Describe the pros and cons of your model. How does it relate to the real world? What do you use it for? Include a graphical representation of your model as well as equations for it.

Reflection: What have you gained by doing this project, both mathematically and otherwise?

Bibliography: Include sources used for the business, the mathematics, and the data. Use the APA (American Psychological Association) citation style.

Journal Questions as used in 2009: To be given out each week. These questions were posted online each week in a discussion forum, so the students could see each other's answers.

Journal 1: *What do you expect from this class?*

Journal 2: *Reflect on the subprime mortgage crisis.*

Journal 3: *What is a reliable source?*

Journal 4: *What is a model? What is a mathematical model?*

Journal 5: *Do you think it is important to have mathematical models for real life problems? Why or why not?*

Journal 6: *What are you doing to prepare for the midterm exam?*

Journal 7: *Why did you choose your particular data set? Was there agreement in your group? Did you want to pick a different one?*

Journal 8: *What could you have done to better prepare for the midterm exam? Was there anything that surprised you?*

Journal 9: *Do you think that mathematics can have a direct impact on society? Explain why or why not.*

Journal 10: *What do you think the government should do (if anything) to regulate the mortgage industry?*

Journal 11: *Do you feel that this class is starting to prepare you for SOA Exam FM/CAS Exam 2? When are you planning on taking the exam? Why then? If you are not planning on taking the exam, why not?*

Journal 12: *Go back and read your first journal. Has this class met your expectations? What did you like about this class? What did you dislike?*

Journal 13: *What have you gained by doing this project, both mathematically and otherwise?*

Journal 14: *What advice would you give to students taking this class next semester?*

Bibliography

[1] Katalina M. Bianco, J.D. (2008) *The Subprime Lending Crisis: Causes and Effects of the Mortgage Meltdown.* Wolters Kluwer Law & Business.

[2] Samuel A. Broverman, *Mathematics of investment and credit*, 5th ed., ACTEX Academic Series, ACTEX Publications, Inc., Winsted, CT, 2010. MR3309182

[3] J.C. Chandor (Director). (2011) *Margin Call* [Motion Picture], United States: Before the Door Pictures, Benaroya Pictures, Washington Square Films, Margin Call Productions, Sakonnet Capital Partners, and Untitled Entertainment.

[4] CNN Money. (2008) Special Report, Issue #1: *America's Money Crisis. Subprime crisis: A timeline.*

[5] Data.gov. (2011). *HECM Single Family Portfolio Snapshot.* Available from `http://explore.data.gov/browse?q=mortgage`, accessed on September 1, 2016.

[6] Adam McKay (Director). (2015). *The Big Short* [Motion picture]. United States: Paramount Pictures.

[7] United States Census Bureau. (2011). *State Median Income.* Retrieved from `http://www.census.gov/hhes/www/income/data/statemedian/index.html`.

15

Using Calculus to Model Income Inequality

Bárbara González-Arévalo and
Wilfredo Urbina-Romero

Abstract. The Gini coefficient (also known as Gini index or Gini ratio) is commonly used as a measure of income or wealth inequality. In this project, students research the Gini coefficient and its applications to the study of income and wealth distribution. To do this, students find a reliable data set, compute the Gini coefficient using integral calculus, analyze the quality of the data, and discuss sources of possible errors. **Keywords:** Income inequality, Gini coefficient, curve fitting techniques, integral calculus, numerical integration, group work.

This project is divided into six parts, each due every two-to-three weeks. It was designed to be used in a second-semester calculus course. Students are therefore typically freshmen or sophomores. The project relies heavily on the notion of area between two curves, so it is mainly an application of the definite integral.

The social justice goal of this module is the study of income inequality through the Gini coefficient, which was originally developed by statistician and sociologist Corrado Gini in 1912 and by definition measures the inequality among values of a frequency distribution. The most well-known application of the Gini index is as a measure of inequities in the income distribution of a nation or a given population. In the following we develop most of the relevant mathematics involved to make sure the material is thoroughly understandable and easy to use.

1 Mathematical Content

The Gini coefficient was developed by the Italian statistician and sociologist Corrado Gini (1884-1965) in his 1912 paper *Variability and Mutability* [4]. Since the definition of the Gini coefficient is based on the Lorenz curve, introduced by Max Otto Lorenz (1876-1959) in 1905 [7], in order to understand the Gini coefficient, one needs to know about the Lorenz curve.

The *Lorenz curve* $y = L(p)$ plots the percentage $L(p)$ of the total income of a population that is cumulatively earned by the bottom $p\%$. In economics, the Lorenz curve is a graphical representation of the cumulative distribution function of the empirical probability distribution of a resource (usually wealth or income) for representing inequality of the resource distribution [13]. The graph shows the proportion of the distribution assumed by the bottom $p\%$ of the values. It is often used to represent income distribution, where it shows for the bottom $p\%$ of households, what percentage $y\%$ of the total income they have. In such use, many economists consider it to be a measure of social inequality. A Lorenz curve is a convex function, always starting at $(0,0)$ and ending at $(1,1)$.

For a population of size N, with a sequence of values $p_j, j = 1, \dots, N$, that are indexed in non-decreasing order ($p_j \leq p_{j+1}$), the Lorenz curve is the continuous piecewise linear function connecting the points $(p_j, L(p_j))$, for $i = 0, \dots, n$, where $p_0 = 0, L(p_0) = 0$, $p_n = 1, L(p_n) = 1$. The *Gini coefficient* is the normalized area between the Lorenz curve $y = L(p)$ and the line $y = p$ of perfect equality (area denoted by **A** in Figure 15.1). That is, it is the ratio between the area **A** and the total area under the curve of equality. Hence, mathematically speaking:

$$G = 2 \int_0^1 (p - L(p))\, dp = 1 - 2 \int_0^1 L(p)\, dp. \tag{15.1}$$

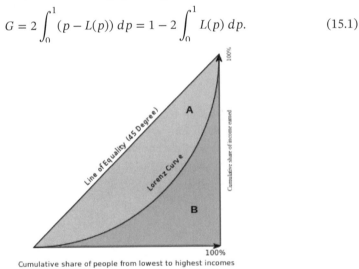

Figure 15.1. Gini coefficient (source: `Wikipedia` [12]).

During this project, students work with all of these concepts, constructing an appropriate Lorenz curve, and calculating the Gini coefficient for a particular data set. To this end, students should be comfortable working with graphs and areas between them. Other than that, the standard preparation for a second-semester calculus should suffice.

For more on the Gini coefficient and the Lorenz curve, see [3] or [6].

2 Context / Background

2.1 Social Justice Context. The Gini coefficient (also known as the Gini index or Gini ratio) is a measure of statistical dispersion intended to represent how equitable a resource is distributed in a population. A Gini coefficient of zero means perfect equality, where all values are the same (for example, where everyone has the same income). A Gini coefficient of one (or 100%) expresses maximal inequality among values (for example, where only one person has all the income). However, a value greater than one may occur if there are negative contributions to the total (for example, having negative income or wealth). For larger groups, values close to or above 1 are very unlikely in practice.

Worldwide, Gini coefficients for income range from approximately 0.23 (Sweden) to 0.70 (Namibia) although not every country has been assessed. Based on Census Bureau data, in 2014 and 2015, the United States has a Gini coefficient of 0.46. See [12] for more information on income inequality in a range of countries as measured by their Gini coefficients.

2.2 Institutional Context. Roosevelt University is an independent, non-profit, metropolitan university with two distinct campuses located in downtown Chicago and suburban Schaumburg, Illinois. With a rich history and progressive curricula Roosevelt University is committed to the highest standards of academic excellence. The faculty and staff take pride in pushing Roosevelt's remarkably diverse students to the limits of achievement, inspiring the transformation of lives and communities through the principles of social justice.

In order to increase student interest in mathematics and its connections to real applications, Roosevelt University began incorporating semester-long projects into its *Calculus II* (integral calculus) course. Beginning in the summer of 2005, Roosevelt University started to send teams of faculty in mathematics and the sciences to the Science Education for New Civic Engagements and Responsibilities (SENCER) Summer Institutes [5]. Faculty members there were awarded a small grant to redesign *Calculus II* to include a project [5] that would accomplish the broad goal of SENCER: to connect science and civic engagement by getting students engaged in "complex, contested, capacious, current, and unresolved public issues" [8]. The Gini index and income inequality theme was a perfect match for SENCER's stated goals.

The issue of wealth and income inequality is the great moral issue of our time, it is the great economic issue of our time, and it is the great political issue of our time. In the United States, income inequality, or the gap between the rich and everyone else, has been growing markedly, by every major statistical measure, for some 30 years. Given the social justice mission of Roosevelt University, providing students with mathematical tools to understand and analyze this problem seems not only convenient but imperative.

3 Instructor Preparation

As explained before, the project consists of six parts. A new part is due every two to three weeks, but each of the first five could be used independently with minor modifications. The sixth part stresses mathematical communication and dissemination, uniting all the previous components.

Below we list a few general planning suggestions for the instructor:

- Since this project is an application of the definite integral, the Fundamental Theorem of Calculus should be covered (at the level of Chapter 5 in [11] for instance) before Part 5 of the project is assigned.

- Numerical integration techniques must also be discussed, in particular the trapezoidal rule (at the level of Section 8.7 in [11] for instance). Also a discussion and review of some mathematical software (Mathematica, Maple, Excel, etc.) will be needed.

- To find the Lorenz curve in Part 4 of the project, students need to study some definitions and concepts from curve fitting and regression analysis. Using the library, their professors, and/or reputable internet/written sources, they need to research the following topics: curve fitting, interpolation, smoothing, and regression analysis (linear regression and non-linear regression); see for instance [2].

3.1 Finding and Working with Data. Given the nature of the problems that the students may choose, it is not always easy to obtain accurate data and in the proper format. Three databases which we recommend are the *U.S. Census Bureau* [10], *C.I.A. World Factbook* [1], and *UNdata* [9]. Students will have to put the data into a proper cumulative format. An estimate of total wealth is also crucial to the project. Economic data are almost always reported in aggregate form.

Usually we obtain tables where one column lists the number of people in a given range of income x_i and another gives the mean income h_i for the group. Then the first step in calculating a Gini coefficient from such a table is to find the points on the Lorenz curve $y = L(p)$. To do this, we need to define the size of the population N and the total wealth T:

$$N = \sum_{i=1}^{n} x_i, \text{ and } T = \sum_{i=1}^{n} x_i h_i.$$

With this notation, the mean amount owned is $\overline{X} = T/N$. The points $p_j = \frac{1}{N} \sum_{i=1}^{j} x_i$, are points along the p-axis between 0 and 1, with $p_0 = 0$. Then the Lorenz curve at p_j is given by

$$L(p_j) = \frac{1}{T} \sum_{i=1}^{n} x_i h_i, \qquad (15.2)$$

the fraction of the total earned by the poorest fraction p_j. With a little algebra, (15.2) can be rewritten as

$$L(p_j) = \sum_{i=1}^{n} \frac{h_i}{T/N}(p_i - p_{i-1}) = \int_{0}^{p_j} s(p)dp, \qquad (15.3)$$

where $s(p) = \frac{h_i}{\overline{X}}$ for $p_{j-1} < p < p_j$ is called the *share density*.

4 The Module

This project consists of six parts, which are detailed below and in the appendix. The following is a suggested timeline:

Project Part	Type of Assignment	Semester System Due Date	Quarter System Due Date
Part 1	Individual	Week 2	Week 2
Part 2	Individual	Week 4	Week 3
Part 3	Groups of 3	Week 7	Week 5
Part 4	Groups of 3	Week 10	Week 7
Part 5	Groups of 3	Week 12	Week 9
Part 6	Groups of 3	Week 14	Week 10

For each of these parts, some class time should be spent explaining to the students what they are expected to do. This should take no more than 5-10 minutes for most parts. Part 1 and Part 4 might need closer to 30-45 minutes. For Part 1 the whole project needs to be explained, along with some basic concepts. For Part 4 a more detailed explanation of what the students need to do will be necessary. This might be a good place to discuss numerical integration techniques. There also needs to be some class time dedicated to the mathematical software (Mathematica, Maple, Excel, etc.) that will be used in the project. The amount of class time will vary greatly depending on students' computational skills.

Part 1. Gini coefficient and Lorenz curve. For Part 1 of the project, students learn about the Gini coefficient and the Lorenz curve, and their applications. In particular, they need to understand how these mathematical constructions are used to describe income inequality. See Appendix A.1 for a possible handout to use in this part.

A visit to the university library should be organized to determine which resources are available to study those notions as well as the possible projects and databases. The availability of reliable data is crucial in order to decide if the project is feasible.

Part 2. Formulating a project using the Gini coefficient. For Part 2 of the project, students propose a project that uses the Gini coefficient. For instance, they could consider studying the evolution of the Gini coefficient in a given country, region, state, or county in a certain year or period of time; they could choose to compare the Gini coefficient and/or evolution of it for two countries or a group of countries, regions, states, or counties, or they could propose to study the distribution of wealth of a given group of a given population. Brief student instructions are provided in Appendix A.2.

For example, the following is a list of projects done at Roosevelt University by students in *Calculus II*:

- *"Gini coefficient in the United States from 1970 until 2010",*

- *"Comparison of wealth distribution in five well-known countries (United States, Italy, Germany, Greece, China)",*

- *"Sweden vs. United States: Lorenz curve and Gini index",*

- *"Brazil's wealth distribution (2004-2009)",*

- *"Gini coefficient of the United States income through four regions (northeast, south-west, southeast, west, midwest)",*

- *"African-American wealth distribution in comparison to the general population",*

- *"A comparison between the richest and the poorest counties in the U.S. (Loudun county, VA vs. Ziebach county, SD)".*

Instructors should emphasize that before students choose their project, they need to make sure that they can find enough reliable data to carry out their analysis.

Part 3. Finding points of the Lorenz curve of the problem. For Part 3 of the project, students work in groups of three, and will keep working in these groups for the remaining parts. They need to decide which of the three different proposed projects (from Part 2) is the one that is going to be the group project. The availability of good quality and reliable data should be heavily weighted in that decision. Then, they need to work with the data in order to put it in the proper (cumulative) format to obtain the points on the Lorenz curve.

Students need to collect a data set of at least twenty-five points; a larger set is even better. They need to make sure that the data are reliable, and they need to discuss this in their project. The quality of the data file created in this part will be crucial for the development of the project throughout the semester.

A brief handout of student instructions to use for this part is provided in Appendix A.3.

Part 4. Curve fitting and regression analysis: The Lorenz curve. Part 4 of the project builds on the data set obtained in Part 3; see Appendix A.4 for a brief handout to be used with this part. It involves fitting the data points obtained in Part 3 to create an approximate graph of the Lorenz curve. In order to make progress towards for-mulating mathematical models for income distribution, students need to study some concepts such as curve fitting and regression analysis. In order to do this, they need to make use of the library, their professors, and/or reputable internet/written sources. Once they learn the basics of curve fitting and regression analysis, they can then find possible approximations of the Lorenz curve. At this point they need to consider whether the data are aggregated or not, the concavity of the data, and whether to use a piece-wise function or not.

Students need to use technology for this part, such as Maple, Mathematica, Matlab, or (for simpler fits) Excel. We strongly recommend that students spend some class time on a curve-fitting computer lab in order to help them with the technology.

Since the Lorenz curve is a distribution function on $[0, 1]$, there are formulas to compute the Lorenz curve associated to a discrete or continuous probability distribu-tion, and ambitious instructors might consider extending the module to incorporate such computations; see Section 5.2.

Part 5. Computing the Gini coefficient. Part 5 of the project builds on Part 4 and involves the actual computation of the Gini coefficient. The final result of this part is the Gini coefficient of the problem for the different approximations of Lorenz curve obtained in Part 4 using formula (15.1) directly or using numerical integration techniques.

A brief handout to be used in this part of the project is included in Appendix A.5.

Part 6. Paper/poster creation. In this sixth and final part of the project, students write a paper or present a poster about their project. We include the handout we use for this part in Appendix A.6.

At Roosevelt University, a poster session is held every semester, and this allows students to ask other groups about their choices. This leads to a rich discussion on the difficulties in finding precise data, and the limitations of modeling.

Because students have a chance to correct their work from Part 5, it is important to provide feedback on Part 5 in a timely manner. An important issue to address is the conclusions that can be drawn from their research. Students need to be able to communicate well; we recommend that they write for an audience who has only taken *Calculus I*. Ideally, students will present their work at a university-wide STEM event.

5 Additional Thoughts

5.1 Application Notes. Starting in Fall 2011, this project has been used several times at Roosevelt University in *MATH 232: Calculus II*. Students in *MATH 232* have already learned about antiderivatives in the previous term; in this course, they begin with Riemann sums and finish the term with sequences and series. The project parts that deal with areas between curves have thus been timed to occur when the class is working on applications of integration.

In practice, most of the project parts are completed in groups of approximately three students. Each student is tested, however, on the contents of the project parts by including an exam question that is similar to a small piece of a project part. This helps the students feel that every individual is accountable for every part of the whole. Students are told this verbally ahead of time, and it seems to help with the group dynamics.

5.2 Possible Variants. There exist formulas to compute the Lorenz curve associated to a probability distribution (both discrete and continuous). For example, the Pareto family of distributions is frequently a good model for wealth distributions. In order to use this approach, Part 3 of the project could be modified to include a brief introduction to probability theory and the notion of a probability distribution function, since most of the students will have not taken an introductory course in probability. In our experience this variant has not been received as positively. In general, students find this extension a significant burden in addition to the material that is already covered. However, it is conceivable that it may work well with more advanced students, maybe in an honors section of the course.

Acknowledgments. We thank the referees for their observations, corrections, and comments. The presentation of the paper was greatly improved thanks to their patient and careful work. The deficiencies of the paper that probably still persist are our own responsibility.

Appendix A: Assignments and Handouts

The following are sample handouts for the students for each part of the project.

A.1 Handout for Part 1. You need to learn about the Gini coefficient, the Lorenz curve, and their applications. In order to do this, you might need to make use of the

library, your professors, and/or reputable internet/written sources. Write down explicitly your answers to the following questions:

(1) *What is the Gini coefficient?*

(2) *What is the Lorenz curve?*

(3) *What is the meaning of the Lorenz curve?*

(4) *How can the Gini coefficient be computed?*

(5) *What is the interpretation of the Gini coefficient?*

A.2 Handout for Part 2. For the second part of the project you will propose a
research question that uses the Gini coefficient. For instance, you could study the evolution of the Gini coefficient in a given country, region, state, or county in a certain
year or period of time. You can compare the Gini coefficient and/or evolution of it for
two countries or a group of countries, regions, states, or counties. Alternatively, you
can study the distribution of wealth of a given group of a given population. Before you
finalize your research question, you must make sure you can find enough reliable data
to carry out the project.

A.3 Handout for Part 3. Economic data are almost always reported in aggregated
form. Usually we find tables where one column lists the number of people in a given
range of income x_i and another gives the mean income h_i for the group.The first step
in calculating a Gini coefficient from such a table is to find the points on the Lorenz
curve $L(p)$. The Lorenz curve can often be represented by a function $y = L(p)$, where p
is represented on the horizontal axis and it is a number on the interval $[0, 1]$, and $L(p)$
is represented on the vertical axis. A Lorenz curve always starts at $(0, 0)$ and ends at
$(1, 1)$.

For a population of size n, with a sequence of values $x_i, i = 1, \dots, n$, that are indexed
in non-decreasing order $(x_i \leq x_{i+1})$, the Lorenz curve is the continuous piecewise
linear function connecting the points $(x_i, L_i), i = 0, \dots, n$, where $x_0 = 0, L_0 = 0, x_n =
1, L_n = 1$. First we need to define the size of the population N and the total wealth T

$$N = \sum_{i=1}^{n} x_i \text{ and } T = \sum_{i=1}^{n} x_i h_i.$$

With this notation the mean amount owned is $\overline{X} = T/N$. The points $p_j = \frac{1}{N} \sum_{i=1}^{j} x_i$,
are points along the p-axis between 0 and 1, with $p_0 = 0$. Then the Lorenz curve at p_j
is given by

$$L(p_j) = \frac{1}{T} \sum_{i=1}^{n} x_i h_i,$$

the fraction of the total earned by the poorest fraction p_j. With a little algebra, $L(p_j)$ can
be rewritten as

$$L(p_j) = \sum_{i=1}^{n} \frac{h_i}{T/N}(p_i - p_{i-1}) = \int_0^{p_j} s(p) dp,$$

where $s(p) = \frac{h_i}{\overline{X}}$ for $p_{j-1} < p < p_j$ is called the *share density*.

You need to collect a data set of at least twenty-five points; a larger set is even
better. You need to make sure that the data are reliable, and if needed, you need to

put your data in a cumulative form as explained above. This file will be crucial for the development of the project throughout the semester.

A.4 Handout for Part 4. You need to study the following concepts of curve fitting and regression analysis using the library, your professors, and/or reputable internet/written sources:

a) Curve fitting

b) Interpolation

c) Smoothing

d) Regression analysis

e) Linear regression

f) Non-linear regression

Once you have learned the basics on curve fitting and regression analysis, you need to construct the Lorenz curve for your data. Here is how you should go about doing this:

(1) *Pick the data sets from Part 3. Try to pick the most complete and reliable one. How reliable do you think your data are and why? Can any real-life data be 100% reliable? What could have led to errors in a particular data set?*

(2) *Use polygonal approximation to obtain an approximate Lorenz curve.*

(3) *Use curve fitting to obtain an approximate Lorenz curve.*

(4) *Use linear and non-linear regression to obtain an approximate Lorenz curve.*

(5) *Compare the approximate Lorenz curves obtained above. What is your conclusion? Which one fits the data set best?*

A.5 Handout for Part 5. In this part of the project you will compute the Gini coefficient for your specific research problem. Below is a series of question to guide you through this process.

(1) *Compare the approximate Lorenz curves obtained in Part 4. Which one is easier to use for computing the Gini coefficient?*

(2) *For each approximate Lorenz curve obtained compute the Gini coefficient, using formula (15.1) directly or using numerical integration.*

(3) *What are your conclusions?*

A.6 Handout for Part 6. The final part of the project requires you to compile your findings into a poster presentation and/or a research paper.

The required elements for the paper/poster are as follows:

(1) *Identifying information:*

    (a) *Title of project;*

    (b) *Names of group members;*

    (c) *University name;*

(d) *Date.*

(2) *Introduction and background on the problem;*

(3) *Description and explanation of data collection, data sources, and methods of handling data;*

(4) *Summary and tables of data used in your model;*

(5) *Presentation of your model with all terms explained;*

(6) *Graphs of the data and your model;*

(7) *Discussion of results and conclusions:*

    (a) *Strengths and advantages of your model;*

    (b) *Limitations and problems with your model.*

(8) *References for any sources used, using the APA (American Psychological Association) citation style.*

Bibliography

[1] Central Intelligence Agency, *The World Factbook*, available at `https://www.cia.gov/library/publications/the-world-factbook/`, accessed on September 2, 2016.

[2] Philip J. Davis, *Interpolation and approximation*, Blaisdell Publishing Co. Ginn and Co. New York-Toronto-London, 1963. MR0157156

[3] Frank A. Farris, *The Gini index and measures of inequality*, Amer. Math. Monthly **117** (2010), no. 10, 851–864, DOI 10.4169/000298910X523344. MR2759359

[4] Gini, C. *Variabilita e mutabilita.* C. Cuppini, Bologna, 156 pages, 1912. Reprinted in *Memorie di metodologica statistica* (Ed. Pizetti E, Salvemini, T). Rome: Libreria Eredi Virgilio Veschi, 1955.

[5] González-Arévalo, B. and Pivarski, M. "The Real-World Connection: Incorporating Semester-Long Projects into Calculus II," *Science Education and Civic Engagement: An International Journal*, Winter 2013. Available online at `https://seceij.net/seceij/winter13/real_world_conn.html`, accessed on September 2, 2016.

[6] Robert T. Jantzen and Klaus Volpert, *On the mathematics of income inequality: splitting the Gini index in two*, Amer. Math. Monthly **119** (2012), no. 10, 824–837, DOI 10.4169/amer.math.monthly.119.10.824. MR2999586

[7] Lorenz, M. O. "Methods of measuring the concentration of wealth," *Publications of the American Statistical Association*, Vol. **9** New Series. no 70 (1905), pages 209–219.

[8] SENCER-National Center for Science and Civic Engagement, *The SENCER Ideals*, 2015. Available online at `http://www.sencer.net/About/sencerideals.cfm`, accessed on September 2, 2016.

[9] United Nations Statistics Division, *UNdata: A World of Information*, 2016. Available at `http://data.un.org/`, accessed on September 2, 2016.

[10] United States Census Bureau, website at `http://www.census.gov/`, accessed on September 2, 2016.

[11] Weir, M. D.; Hass, J. and Giordano, F. *Thomas' Calculus*, Eleventh Edition, Addison-Wesley Longman Publishing Co., Inc. Boston, MA, USA, 2005.

[12] Wikipedia, *Gini coefficient*, available online at `http://en.wikipedia.org/wiki/Gini_coefficient`, accessed on September 2, 2016.

[13] Wikipedia, *Lorenz curve*, available online at `http://en.wikipedia.org/wiki/Lorenz_curve`, accessed on September 2, 2016.

16

What Does *Fair* Mean?

Kira Hamman

Abstract. How votes are counted is a critical but largely ignored variable in the democratic process. The default method is usually plurality, which has the benefit of being easy to understand and implement. Unfortunately, it is also usually the worst possible method in terms of fairness, whether *fair* is defined as choosing the candidate that most voters prefer, or as *not* choosing a candidate that most voters dislike. In this module students investigate the problems with plurality voting and explore alternative voting methods that avoid those problems. In particular, we introduce preference ballots and two different ways of tabulating results based on them: instant runoff voting and the Borda count. Students begin by implementing the methods using calculators; if desired, they can go on to implement them using spreadsheet software in the optional project. Finally, we discuss Arrow's Impossibility Theorem and its consequences. Students come away from the module with a deeper understanding of the relationship between fair elections and the mathematics of voting. **Keywords:** Instant runoff voting, Borda count, Arrow's Impossibility Theorem, democracy, voting, fairness, quantitative literacy, inquiry-based learning, spreadsheets.

Nearly 200 years ago, Alexis de Tocqueville declared that "democratic peoples show a more ardent and more lasting love for equality than for freedom". While a quick look at the current income distribution in this country provides a compelling counterexample to this statement, if we replace the word *equality* with the word *fairness* we have a much more plausible claim. In our democracy we strive for fairness in education, employment, and housing; we believe in fair trade and equal opportunity. Perhaps most importantly, we work to have fair elections.

Yet the question of what constitutes fairness is deeply problematic in the context of the mathematics that underlies the democratic process. From how we count the vote, to how we apportion Congress, to how we structure the United Nations, what is *fair* is often — indeed, usually — not obvious. Is plurality voting fair? Is the Electoral College fair? What about veto power? Should we choose an apportionment method that favors small states or large ones? What does *fair* mean in a democratic society?

In this module, students are asked to think about that question in the context of what is arguably the most fundamental piece of a democracy: voting. They are given social and historical context and taught the mathematics necessary to make the relevant calculations. As they work through this topic, they come to understand both the concept of fairness and the intricacies of democracy in a new way.[1]

1 Mathematical Content

The content of this module is appropriate for a variety of social science courses, including courses on political science, government, and history. In a political science or government class, the material might be used to supplement a discussion of the branches of government, the Constitution, Congress, apportionment, or districting and redistricting. In history classes, particularly United States history classes, an instructor could play up the material on historical context and look at the impact historical understandings of fairness have had on our democracy today. And, of course, the module can be used in any quantitative literacy class that is topics-based. That is, in a class structured as a series of mathematical topics, this module can be inserted as one topic.

Very little existing mathematical knowledge is required of students to understand this material; arithmetic and beginning algebra will suffice. Some parts of the module take that basic mathematics and use it in new and different ways, so that students acquire the ability to use the mathematics they already know more flexibly. Other parts of the module provide opportunities for students to think analytically and contextually about the mathematics they are doing, rather than to simply learn and implement procedures. In both cases, they will need to understand and organize quantitative information for a purpose other than obtaining a single expected answer. This experience of mathematics in action provides a powerful antidote to the "when will I use this?" refrain so common in lower-level mathematics classes.

The principal mathematical topics covered in the module are plurality voting, preference ballots, instant runoff voting, the Borda count, and Arrow's Impossibility Theorem. We offer a brief discussion of each of these below.

1.1 Plurality Voting. Plurality is the most commonly used voting method in the world because it is the simplest. Often referred to as *winner take all*, plurality uses a single vote from each voter — a vote for that voter's favorite candidate — and then declares the candidate with the most votes the winner. Most elections in the United States are run this way, including presidential and congressional elections. If such an election has only two candidates, and if we ignore the possibility of a tie, then one of

---

[1]EDITOR'S NOTE: See Chapter 11 (*Voting with Partially-Ordered Preferences* by John Cullinan and Samuel Hsiao) for more on voting theory.

the candidates will necessarily receive more than half of the votes. That candidate is preferred by a majority of voters and can therefore reasonably, and fairly, be considered the winner. However, as soon as a third candidate is introduced we can no longer guarantee the existence of a majority winner, and the situation becomes immensely more complicated. The module explores the difference between majority and plurality winners and examines the problems with plurality voting.

1.2 Preference Ballots. The root of the problem we see with plurality is that only a voter's first choice matters. There is an alternative to this, called *ranked choice voting* or *preference ballots*, where voters are asked not just for their first choice candidate but to rank all the candidates in order of preference. Australia, India, and Ireland all use preference ballots in their national elections, as do many states and cities both in this country and around the world. How many different preference ballots are possible is a basic combinatorial question, and offers an opening to talk about that branch of mathematics if desired. Once preference ballots have been cast, any one of a variety of methods can be used to tally the votes. These methods avoid many of the problems that come up with plurality voting, although it will turn out to be impossible to avoid all undesirable situations with any single method. The module covers two such methods in detail: instant runoff voting and the Borda count.

1.3 Instant Runoff Voting. Instant runoff voting (IRV), sometimes called single transferable vote, is a method that is used in this country by some state and municipal elections and in a number of other countries around the world for national elections. To implement IRV, rather than looking at who has the *most* first place votes and calling that candidate the winner, we look at who has the *fewest* first place votes; eliminate that candidate; and redistribute the votes originally cast for that candidate among the remaining ones, by tallying each of the next-highest ranked candidates on those ballots. We repeat this process until only two candidates are left; one of those will necessarily be a majority winner. IRV avoids situations in which a "spoiler" third candidate can pull a few votes from a leading candidate who would otherwise have a majority, giving the victory to a challenger who has less than 50% of the vote but more than the leader, once the spoiler is taken into account. Think Bush, Gore, and Nader in 2000 — a situation that would not have happened using IRV.

1.4 Borda Count. Another way to tally preference ballots is with what is known as a Borda count, named for Jean-Charles de Borda, the French scientist who invented it. In a Borda count, points are assigned for each ranking on the ballot. Typically, last place earns zero points, second-to-last place earns one point, and so on up to $n-1$ points for first place, where n is the total number of candidates. Points are awarded for each ballot and are summed to get a total number of points for each candidate. The candidate with the most points wins. The Borda count favors candidates who no one has strong negative feelings about, and is therefore sometimes used in places with a high level of political strife, or by organizations seeking to elect people who are acceptable to a broad group.

1.5 Arrow's Impossibility Theorem. The module concludes with a discussion of Arrow's Impossibility Theorem. In 1950, economist Kenneth Arrow set out criteria for fairness in the context of voting and elections. The criteria can be stated formally in a number of different ways, but informally they say that (a) if all voters prefer candidate A to candidate B, then candidate B should not win, and (b) this should be true regardless of voters' preferences for other candidates. These seem like pretty reasonable criteria. Nevertheless, Arrow proved that when there are more than two candidates, it is impossible to have a voting system that always satisfies them. Actually, that is not quite true. There is one voting system that works: dictatorship, the system in which only one person's vote counts. But we can probably agree that dictatorship is not fair!

More precisely (but still not very formally), Arrow's Theorem can be stated as follows:

> **Arrow's Theorem:** *In an election with more than two candidates, no voting method can ever satisfy all of the following conditions:*
>
> - *it always produces a winner,*
> - *it is not a dictatorship,*
> - *if all voters prefer candidate A to candidate B, then candidate B does not win, and*
> - *if A is the winner, a voter changing her ballot will not cause B to win instead unless the change puts B ahead of A.*

Arrow's Theorem will be many students' first exposure to a mathematical result with "real life" implications. In addition, it provides an opening to discuss the notion that mathematics can be used not only to figure out how to do things, but also to see that some things can never be done.

2 Context / Background

Fifty years ago, political awareness among college students was a given. Civil rights, feminism, the war in Vietnam — college campuses were alive with discussion of, and action on, these issues. After decades of dormancy, today this political awareness seems to be returning, albeit in a somewhat different form. The 2008 Presidential election saw a significant increase in youth activism, and the 2016 election cycle took that involvement to new highs — and lows. Many students have no faith in a political process they perceive to be at best ineffectual and at worst corrupt. And, in fact, it is difficult to counter that perception. The system *is* frequently ineffectual, or worse. What our students do not yet understand, and what we would do well to communicate to them, is that most of these problems are well understood, and many have solutions. But correcting the problems will require an educated electorate willing to work for change. If we are to have that, we must explicitly teach this material.

The real purpose of this module, and probably all the modules in this book, is to empower students to work for social justice through a democratic process that they understand. Even better, to inspire students to create change that will make the process itself more socially just. The mathematics (to say nothing of the social science and thousands of years of history) makes the limitations of democracy very clear, but it also makes clear the places where improvement is possible. With an understanding of the underlying principles, the flaws in the system become less overwhelming. Time and again students come to this material apathetic and dispirited about government and leave inspired to work for change. For an educator, that is a win.

To get the most out of this module, students should come to it with a basic understanding of how voting works in the United States, and of some of the historical problems with it. In particular, reference is made to the 2000 Presidential election. A broad sense of global politics is also helpful, as it will allow students to understand the connection between a country's voting method and its political situation. With all that said, the module itself can serve as a springboard for discussion of those topics. Students lacking background knowledge of politics and history will still be able to understand this material.

3 Instructor Preparation

For the module to be successful, the instructor must have an understanding of both the mathematical and the philosophical issues in question. The mathematics of voting theory, and social choice theory more broadly, can be found in many quantitative literacy textbooks. Chapter 9 of *For All Practical Purposes* [2], for example, provides a good introduction to the topic and the voting methods covered in the module.

Philosophical preparation is less straightforward. Poundstone's *Gaming the Vote: Why elections aren't fair and what we can do about it* [3] and Saari's *Chaotic Elections: A mathematician looks at voting* [4] both provide excellent discussions of the issues, and are engaging for mathematicians and non-mathematicians alike. Of course, an entire book (or two!) may be a bit much preparation for a single module. Therefore, for a general overview of the issues, we suggest chapter 9 of Poundstone and the beginning of chapter 1 in Saari. For the historian, chapter 3 of Poundstone offers a concise and engaging discussion of the history of the problem. The mathematically inclined political scientist might try chapter 6 of Poundstone, although such a person will undoubtedly be fascinated by this material and end up reading all of both books. And of course the truly interested can read Arrow's own treatment of the topic [1].

Ultimately, instructors will need to think through the questions in the module themselves, and develop their own thoughts and opinions. What *does* fair mean? Of course, there is no correct answer, but there is a big difference between a well-reasoned opinion supported by evidence from the mathematics, and a guess. These are complex, difficult questions, and a thoughtful discussion of them requires a depth of understanding that takes time to develop. The module will work better if the instructor has spent that time and developed that understanding.

The logistical preparation for the module is fairly simple. Students will need calculators, and, if the instructor plans to use the optional project, access to computers with Excel or other spreadsheet software. Instructors should also make sure to understand how to implement the three methods that are discussed, and should work through the project to get a sense of both how long it will take students and what the sticky points might be for their particular groups.

4 The Module

The module fits within one class period, and the time spent on discussion can be expanded or contracted as time and interest permit. Should an instructor wish to spend several classes on this material, additional voting methods could be introduced or additional examples could be explored. See Section 5 for some ideas on incorporating other mathematical topics into the discussion of fairness.

Ideally the module should be used at the point where the topic naturally arises in the context of other course material. This will allow students to contextualize their understanding of the mathematics and will both enrich the discussion of the meaning of fairness and improve the understanding of the significance of the mathematics. The theme of fairness is threaded through the lesson and should be kept in mind throughout. We repeatedly return students' attention to the question *what is fair?*

4.1 Preliminary Discussion. Before beginning a discussion of the question "what is fair?" it may be helpful to spend some time considering what students already know or believe about the concept of fairness. The instructor may begin by asking what they think of when they hear the word "fair". What makes something fair or unfair? What are some examples of each? Offer hypothetical situations for discussion. Is it fair for one sibling to get the larger cookie? What if that sibling is the one who baked the cookies? Is it fair for one school district to have more money than another? What if the additional money comes from donations from parents? What if the district with less money has higher poverty, such that the parents cannot afford to make donations? Then the class may explore examples of fairness questions that do not involve issues of more and less. Are mandatory minimum sentences fair? Is it ever fair to prosecute a juvenile as an adult? Are all-male or all-white governing bodies fair? Finally, it will help to consider what makes a situation fair or unfair. What criteria are we implicitly using to make this judgment?

4.2 Plurality Voting. The instructor should introduce plurality voting by considering the following example:

In a Student Government election with three candidates, the results are as follows:

Candidate A: 400 votes,
Candidate B: 350 votes,
Candidate C: 250 votes.

Most people would declare A the winner of this election, because she has more votes than either of the other two candidates. Candidate A is the plurality winner.

But wait! Suppose A is an extremist candidate — let us say she favors eliminating all sports teams on campus. Furthermore, there are a lot of athletes at this college. In fact, everyone who voted for either B or C is an athlete, so we can assume that candidate A is their last choice. Thus a large majority (60%) of voters would rather have anyone other than A, yet A won the election.

This is the problem with plurality voting, and it happens in real life too. It is what happened in Florida during the 2000 Presidential election, arguably changing the course of our country's history. More on that later.

Additional examples of plurality elections are easy to manufacture if desired. The instructor can stimulate discussion with questions such as: *Is this fair? What problems might it create for the campus? Who will be happy with this result, and who unhappy? What might be done to make the result more fair?*

4.3 Preference Ballots. Next the instructor should ask students to imagine that the Student Government election above had used preference ballots. There are six possible ways to order the candidates on such a ballot:

(1) A B C

(2) A C B

(3) B A C

(4) B C A

(5) C A B

(6) C B A

This is a good place for an optional combinatorics digression. Students can discuss: *Why are there six possibilities? How many would there be with four candidates? List them. How many would there be with ten candidates? (Do not try to list them!) What is the pattern?* [Answers: With three candidates, there are three choices for each ballot's first place. Once the first place is filled, there are two choices left for the second place. Once the second place is filled, the third place is determined. Thus there are $3 \times 2 \times 1 = 6$ possible permutations (orderings) of the three candidates. Therefore, with four candidates there are $4 \times 3 \times 2 \times 1 = 24$ possibilities. In general, with n candidates there are $n!$ (n factorial) possible ballots. A tree diagram is helpful for explaining this.]

Since we have already said that voters who choose either B or C first will always choose A last, we can assume that there are no ballots like (3) or (5). The voters who chose B in the first election will all vote BCA (4), and the voters who chose C in the first election will all vote CBA (6). Therefore we have:

$$\text{B C A: } 350 \qquad\qquad \text{C B A: } 250.$$

We do not know, of course, who the voters who put A first will put second, so let us assume that half prefer B and the other half prefer C. Thus we have:

$$\text{A B C: } 200 \qquad\qquad \text{A C B: } 200.$$

At this point the instructor should ask students to discuss questions like: *What should we do now? How can this information be used to find a winner?* Students should be urged to keep the idea of fairness in mind as they discuss various options.

4.4 Instant Runoff Voting. Now we will use the example above to explain the idea of eliminating the lowest vote-getters and recalculating. In this case, C is eliminated first because she has only 250 first place votes, while B has 350 and A has 400.

Once we eliminate C from this election, what is left? Let us look:

$$\text{B C̶ A (becomes B A): } 350 \qquad\qquad \text{C̶ B A (becomes B A): } 250$$
$$\text{A B C̶ (becomes A B): } 200 \qquad\qquad \text{A C̶ B (becomes A B): } 200.$$

With C out, $350 + 250 = 600$ voters prefer B, and $200 + 200 = 400$ voters prefer A, so B wins the election.

Next we try an example with four candidates. If the class did the optional combinatorics section, they know that there are $4 \times 3 \times 2 \times 1 = 24$ possible ballots. If not, then the instructor can briefly explain this.

Let us suppose that in this situation the candidates' political views range from far left (candidate A) to far right (candidate D). In such a situation, many voters will prefer a moderate candidate (B or C), but those who prefer an extreme candidate will probably rank the rest in order of their politics. That is, voters who prefer A will rank the candidates ABCD (from left to right), and voters who prefer D will rank the candidates DCBA (from right to left). Suppose, then, that we have the following ballots:

A B C D: 90   B C A D: 449

D C B A: 10   C B D A: 451.

Clearly, most voters are moderate, and the majority of the first place votes are split almost evenly between B and C. First, students should consider who would win this election if it were done using plurality. The key to doing this is to ignore everything except the first-place votes, so that what you see is:

A: 90   B: 449

D: 10   C: 451.

In other words, C wins, but only by a hair.

Next students should get into groups of two or three and do an instant runoff vote with these ballots. It should work out as follows:

D is out first, with only 10 first-place votes. With D out, the ballot totals are:

A B C: 90   B C A: 449

C B A: 10   C B A: 451.

At this point, A has 90 first-place votes, B has 449, and C has $10 + 451 = 461$. Thus A is out next, leaving:

B C: $90 + 449 = 539$ C B: $10 + 451 = 461$.

B has a majority, and is the winner.

This outcome generates many possible questions for discussion, including: *Is this a fairer outcome than using plurality? Why or why not? Does this method have its own drawbacks? What might they be? For political science classes: why did we choose the four ballot rankings we did, given the political assumptions we are making? Would you make different assumptions, or different choices?*

4.5 Borda Count. Using the last example with four candidates, we can introduce the Borda count.

A B C D: 90   B C A D: 449

D C B A: 10   C B D A: 451.

Last place earns zero points and we count up from there, so in this election first place would be worth three points, second place worth two points, and third place worth one point. Candidates get points for each ballot cast, and whoever has the most points wins the election.

In our example, candidate A earns points as follows:

- three points for each of the 90 ballots on which she is listed first: $3 \times 90 = 270$ points,

- two points for 0 ballots because no one has listed her second: $2 \times 0 = 0$ points,

- one point for each of the 449 ballots on which she is listed third: $1 \times 449 = 449$,

- zero points for the remaining ballots because she is listed last: $0 \times 461 = 0$.

Thus candidate A's point total is $270 + 449 = 719$.

Then students should calculate point totals for B, C, and D:

$$B : (3 \times 449) + (2 \times 541) + (1 \times 10) + (0 \times 0) = 1347 + 1082 + 10 + 0 = 2439.$$
$$C : (3 \times 451) + (2 \times 459) + (1 \times 90) + (0 \times 0) = 1353 + 918 + 90 + 0 = 2361.$$
$$D : (3 \times 10) + (2 \times 0) + (1 \times 451) + (0 \times 539) = 30 + 0 + 451 + 0 = 481.$$

Thus the Borda count names B as the winner, just as IRV did.

This is essentially what happened in Florida in the 2000 Presidential election. If we think of candidates A, B, C, and D from left to right politically as Ralph Nader, Al Gore, George W. Bush, and Pat Buchanan, we can see that, in a plurality election, the first place votes for Nader effectively "spoiled" the election for Gore. That is, since only first choices mattered, the fact that Nader voters would have preferred Gore to Bush was irrelevant. Enough voters chose Nader over Gore to bring Gore's first place total slightly below Bush's, and Bush won Florida even though a majority of voters (53.9%) preferred Gore to Bush. (We ignore the complication of Florida ballots which may have been mis-marked by voters and simply consider their choices as marked.)

Questions to discuss with students include: *What is the fair thing to do? How should we vote, and how should we count votes, so that elections are as fair as possible? What does* fair *mean in this context? Is it fair for a candidate to win when a majority of the voters prefer a different candidate? Why did Florida matter in the 2000 Presidential election?*

4.6 Arrow's Impossibility Theorem.

At this point, students will be wondering which method is "best". It is time to talk about Arrow's Theorem! Recall the statement of the theorem:

> **Arrow's Theorem:** *In an election with more than two candidates, no voting method can ever satisfy all of the following conditions:*
>
> - *it always produces a winner,*
> - *it is not a dictatorship,*
> - *if all voters prefer candidate A to candidate B, then candidate B does not win, and*
> - *if A is the winner, a voter changing her ballot will not cause B to win instead unless the change puts B ahead of A.*

There is no need to give a formal mathematical statement of the theorem, but the instructor should make sure students understand why each criterion is desirable, and that Arrow's Theorem doesn't just say that we do not have such a system available to us. It says such a system is impossible — that is, we can demonstrate mathematically that such a system will never exist. There is no point wasting our time trying to find one. Such is the power of mathematics!

If we accept Arrow's fairness criteria, this is a pretty depressing state of affairs. It tells us that we will never have a perfectly fair way to vote — that such a system is impossible. But as Arrow himself said, the theorem "is not a completely destructive or negative feature any more than the second law of thermodynamics means that people do not work on improving the efficiency of engines. We're told you'll never get 100% efficient engines. That is a fact — and a law. It doesn't mean you wouldn't like to go from 40% to 50%" [2]. In other words, even if perfect fairness is impossible, we can, and should, strive to make our system fairer than it currently is.

Arrow's Theorem raises all kinds of interesting questions, including: *Do you agree with Arrow's fairness criteria? What other criteria might we set for fairness in voting? Are some of them more important than others? What does it mean for one system to be* fairer *than another?*

4.7 Optional Spreadsheet Project. This project, which uses Excel or other spreadsheet software, gives students a sense of how different voting methods can produce different results. See Appendix A.1 for a handout. Questions to discuss with students in relation to the project include: *What issues are raised by the results of this project? Does this change your thoughts on the question* what is fair?

5 Additional Thoughts

The discussion of fairness can be enriched by the study of additional topics both in and outside of mathematics. Three compelling mathematical topics for extending the conversation are apportionment, which is covered elsewhere in this book,[2] weighted voting, and gerrymandering.

5.1 Apportionment. Apportionment assigns members of one set (acres of land, tax dollars, members of Congress) to members of another, smaller set (farmers, government agencies, states) in proportion to their sizes. That is, the farmer who pays less gets less land, the smaller government agency gets less funding, and small states have fewer Representatives in the House than large states. The problem of apportionment is one of rounding off — we know how to find the right (meaning fair) number of Representatives per state, but what do we do when that number turns out to be 8.5? You can't have half a Congressperson! Should such a state get 8 Representatives or 9? Remember that we have to end up with the correct total number, so rounding one state's number up effectively means you'll have to round another one down, and states get a little grouchy when their numbers get rounded down. Furthermore, we need to be able to deal with changes to population and to the number of members of one set or the other (new states, for example, or less tax money). It turns out that these situations lead to paradoxes similar to the problems with voting methods. That is, criteria that seem intuitively to be fair are violated by certain apportionment methods, and we can prove that no apportionment method will ever meet all the desired criteria. Thus the question *what is fair?* is deeply meaningful in this context.

5.2 Weighted Voting. Weighted voting is a fascinating (albeit mathematically challenging) topic. In a weighted voting system, some votes count more than others or, equivalently, some voters have more than one vote. A corporation in which shareholders have one vote for each share they own is an example of a weighted voting system.

---

[2]EDITOR'S NOTE: See Chapter 9 (*Evaluating Fairness in Electoral Districting* by Geoffrey Buhl and Sean Q Kelly).

Another is the Electoral College, in which each state has a number of votes proportional to its population (a good apportionment problem!) and, with a couple of exceptions, casts them all for the same candidate. A dictatorship, in which one person has all the votes, is an extreme example. It is probably not difficult to see that weighted voting raises all kinds of fairness issues.

5.3 Gerrymandering. Gerrymandering, or drawing districts for political advantage, is an issue which has been much in the news, as courts have weighed challenges to districting in states such as Maryland, North Carolina, and Wisconsin. Many agree that gerrymandering is not fair, but there isn't agreement on how to prevent it. How to draw fair districts is a complex question, the mathematics of which is largely beyond the scope of this module. Nevertheless, districting, redistricting, and gerrymandering all make for interesting and timely extensions of this topic. A quick Google search will turn up a variety of good videos explaining gerrymandering; for a more involved activity try the Annenberg Center's redistricting game, available at `http://www.redistrictinggame.org/`.

5.4 Further Extensions. Instructors who wish to expand the discussion of voting methods can easily make connections to current events. The state of Maine approved a ballot measure to move to preference ballots in statewide elections as of June 2018; as time goes on, more states and municipalities in the United States may follow. How will those moves play out in those places? What issues will come up, either anticipated or unexpected, and in what senses will the new system be an improvement? Coverage in the news media of these questions will no doubt provide excellent material for classroom discussion in upcoming elections.

In addition, there are many ways to extend the fairness conversation beyond mathematics. Equitable taxation; resource distribution; questions of access and opportunity versus outcome in education, health care, and housing; all these and many others offer opportunities for discussions of fairness and unfairness in our society.

The question *what is fair?* is at the heart of our democracy, and we have a responsibility to provide students with the tools to meaningfully grapple with it. Mathematics is one such tool, and is necessary for both true understanding of the problems and the development of real solutions.

Appendix A: Assignments and Handouts

A.1 Optional Spreadsheet Project. A municipal election has four candidates and nine million voters. When the preference ballots come in, the votes are:

3 million: A D B C
1 million: A B C D
1 million: B C D A
1 million: B C A D
1 million: C B D A
1 million: C D B A
1 million: D C B A.

(1) Who is the winner of this election if the plurality method is used? (ANSWER: A)

(2) Who is the winner of this election if instant runoff voting is used? (ANSWER: C)

Set up a spreadsheet to calculate the winner using a Borda count, as follows:

(1) Across the first row of the spreadsheet, type the word "points" and then the point values of the positions on the ballot.

(2) Under the word "points" type the word "votes" and enter the number of ballots of each type, in millions.

(3) Finally, enter the preference ballot orderings. Your spreadsheet should look like this:

Points	3	2	1	0
Votes				
3	A	D	B	C
1	A	B	C	D
1	B	C	D	A
1	B	C	A	D
1	C	B	D	A
1	C	D	B	A
1	D	C	B	A

(4) Under the table on your spreadsheet, enter each candidate's letter (A, B, C, and D) in column A.

(5) Next to the letter A, enter the formula to calculate the Borda score for candidate A. Here is how:

   (i). We want three points for each of the three (million) votes that A got on the first ballot type. To get that, we need to multiply the three in cell **A3** by the three in cell **B1**. Anytime we want Excel to do math, we have to start by typing an equal sign. Thus the formula we want in the cell begins =A3*B1.

   (ii). Next we want to add three points for each of the one (million) votes A got on the second ballot type. The formula for this is A4*B1. We want to add that to the previous product, which gives us =(A3*B1)+(A4*B1).

   (iii). In the next ballot type, A is in last place and doesn't get any points, so we can skip that one.

   (iv). The fourth ballot type has A in third place, which is worth one point for each vote. If we add that product to what we have so far, we end up with

$$=(A3*B1)+(A4*B1)+(A6*D1).$$

   Notice that we're using **D1** because that is the cell that tells us the third-place votes are worth one point each.

   (v). Since A is in last place on the remaining three ballot types, we do not have to add anything else. Hit enter, and the Borda score for A will be calculated in the cell where you entered the formula.

   (vi). Follow this procedure to calculate the Borda scores for B, C, and D.

3. Who is the winner of this election if a Borda count is used? (ANSWER: B)

What issues are raised by the results of this project? Does this change your thoughts on the question: *what is fair?*

Bibliography

[1] Kenneth J. Arrow, *Social Choice and Individual Values*, Cowles Commission Monograph No. 12, John Wiley & Sons, Inc., New York, N. Y.; Chapman & Hall, Ltd., London, 1951. MR0039976

[2] Consortium for Mathematics and Its Applications, *For All Practical Purposes: Mathematical literacy in today's world* (*10th ed.*), New York, W.H. Freeman and Co., 2016.

[3] William Poundstone, *Gaming the vote*, Hill and Wang, New York, 2008. Why elections aren't fair (and what we can do about it). MR2682293

[4] Donald G. Saari, *Chaotic elections!*, American Mathematical Society, Providence, RI, 2001. A mathematician looks at voting. MR1822218

17

Social and Environmental Justice Impacts of Industrial Agriculture

Amy Henderson and
Emek Köse

Abstract. The environmental impacts associated with concentrated live-stock waste are well suited to be studied mathematically, and we provide ordinary differential equation (ODE) models addressing the contribution of excess nutrients to the creation of aquatic dead zones. The mathematics employed are standard, and will fit well into differential equations or mathematical modeling courses. A complete worksheet and fully developed solutions are provided. Through engagement with this activity, students will deepen their understanding of ODEs, which are one of the most powerful tools for modeling real-world problems. We provide extensive supplementary resources which can facilitate a multidimensional comprehensive module spanning multiple weeks. **Keywords:** Industrial agriculture, environmental justice, differential equations, non-linear systems, group work, active learning.

Mathematical engagement can be motivated by compelling social justice issues. Providing students with the opportunity to grapple with concrete information from which they can construct meaning is associated with positive learning outcomes [18].

The learning theory literature also indicates that this inclination may be more pronounced within some groups, such as women and people of color, who are reported to learn better through concrete experiences and active experimentation than through abstract conceptualization [1]. Incorporating mathematical methods which are grounded in concrete social justice problems can thus effectively contribute to attracting and retaining women, minority, and first generation students.

We describe a module informed by a current social justice issue that can be used in any standard differential equations course that covers systems of differential equations, within two to three hours time. More specifically, this module focuses on the environmental contamination caused by industrial livestock agriculture. This serious environmental issue is also a compelling social justice issue, as low-income and minority residents are disproportionately affected by pollution generated from industrial livestock agriculture. The focus here is on a particular example of industrial livestock agriculture: high intensity hog production in North Carolina.

In the module, mathematical modeling allows us to comprehend the nature and extent of the impact of aquatic contamination. In the extended data handling & analysis segment, inferential statistical analysis is used to determine whether the negative impacts observed can be appropriately described as an environmental justice issue. Thus students can see clearly how mathematics can illuminate a contemporary social justice issue.

1 Mathematical Content

We present a differential equations module that is well motivated by the environmental justice issue associated with the noxious pollution generated by industrial hog farms. Students should have successfully completed *Calculus* I and II, and ideally *Linear Algebra*, and be familiar with systems of differential equations. These requirements may be standard prerequisites for *Differential Equations*. Instructors of *Mathematical Modeling* should ensure that their students have sufficient background in differential equations in order to effectively utilize these activities.

This module aims to enable students to translate a story motivated by real-life phenomena into mathematics. The nature of the problem requires interactions between state variables, which yields a near-realistic model with non-linear systems of differential equations. Because differential equations are foundational, it is important that students be exposed to a wide range of applications in order to deepen their mastery of the techniques, and to develop an appreciation of not only their practical relevance but also the power of mathematics to affect change in the world. Analysis of the model leads students to experiment and draw conclusions allowing them to experience the benefits of building and testing their own mathematical models. The module is designed to encourage and require critical thinking.

2 Context / Background

The techniques and materials which form the basis for this module were developed at St. Mary's College of Maryland, a public liberal arts college. St. Mary's College of Maryland is the state of Maryland's public honors college, has an enrollment of approximately 1900 students, and is primarily a residential institution. Upper-level mathematics courses at St. Mary's typically have an enrollment of 20 to 30 students, and professors teach three 4-credit courses per semester. The materials and techniques discussed below have been successfully implemented in this context; we anticipate that they could be used in a wide range of settings, including larger classes or recitation sections.

The social justice issue which motivates this module is industrial livestock pollution. Historically, livestock rearing, to supply meat to consumer markets, was performed primarily by farmers who produced a variety of agricultural products, such as grain and hay to feed their livestock, and sometimes fruits and vegetables as well. These diversified farms were, in many respects, sustainable, as the nutrients in the animals' manure were "recycled" back onto the fields producing grain and hay, returning the nutrients that had been used up in growing the crops back to the soil. With the decline of the diversified family farmer and the rise of industrial agriculture, this symbiotic relationship no longer holds. Industrial livestock production concentrates many animals at a single location, a location that generally cannot produce enough grain and hay to support the livestock. As a result, the feed for the livestock is grown elsewhere—often many states away—and the manure that is generated by the livestock cannot be returned to the soil that produced their food. The manure then becomes pure waste, rather than part of a productive cycle [7].

Concerns regarding industrial livestock production are not limited to those of environmental activists. Researchers at the Government Accountability Office (GAO) reported that industrial livestock production is becoming ever more concentrated with significant negative impacts on both the environment and human health [6]. The GAO report cites numerous studies which have "directly linked air and water pollutants from animal waste to specific health or environmental impacts". GAO found that the Environmental Protection Agency, the agency charged with regulating concentrated animal feeding operations (CAFOs), does not adequately monitor or regulate CAFOs, and that more stringent monitoring and regulation must be enacted in order to safeguard water, air quality, and human health [6].

Our module is motivated by concerns about the far-reaching negative environmental and human health effects of industrial livestock production. Examples of these negative impacts include: dead zones in the Gulf of Mexico and Chesapeake Bay, attributable in large part to excess nutrients associated with animal waste [9]; fish kills and contamination of surface waterways due to manure spills [8]; elevated spontaneous abortion rates in Indiana due to increased nitrate levels in drinking water (groundwater contamination) [2]; increased respiratory disease in populations living near concentrated cattle (often dairy) or hog facilities due to elevated ammonia concentrations in the air; and more [8].

To focus on the issue, we provide resources and data targeted to a specific case: industrial hog production in North Carolina. This industry has grown dramatically over the past several decades in North Carolina, and is concentrated largely in lower-income areas of the state. The industrial-scale hog production facilities rely on "manure lagoons" to contain the tremendous volume of animal waste produced—some individual facilities can produce as much waste in a year as a mid-sized American city—and such lagoons are subject to leaks and breaches. Though we have made tremendous strides in treating human waste in the United States, we have done virtually nothing to treat animal waste despite the fact that such waste generally contains more contaminants—including pathogens like *Salmonella* and *Escherichia coli* (or *E. Coli*) as well as antibiotic residues and heavy metals—than human waste [8].

3 Instructor Preparation

Developing interest: Background reading accessible to all audiences. For classes in which the instructor is primarily using the environmental justice issues associated with industrial-scale hog production, to motivate engagement with the mathematical or statistical methods presented here, the 2006 article "Boss Hog: The Dark Side of America's Top Pork Producer" [14] provides a compelling overview of the subject in a format which is accessible to all audiences. This article covers many of the negative externalities associated with industrial pork production as it explores the increasing concentration of the industry. Students may be interested to learn that since the time [14] was written, Smithfield Foods Inc., the producer it mainly focused on, has been purchased by a Chinese corporation, a transaction that was controversial at the time [13].

"CAFOs and Environmental Justice: The Case of North Carolina," published in the June 2013 issue of *Environmental Health Perspectives* (EHP) is a more rigorous article which extensively cites academic sources, yet remains accessible to most readers. EHP publishes both articles which are academic in format and "news" pieces which are designed to be more accessible to a lay audience. This piece falls in the latter category. This article includes pictures of industrial swine operations, including aerial views of manure lagoons, and also maps which illustrate the relationship between the siting of CAFOs in North Carolina and concentrations of poor and minority populations [8].

Research: Context for the mathematical models. "Addressing externalities from swine production to reduce public health and environmental impacts" [11] provides additional information regarding the human health impacts of industrial swine production. This article would help contextualize the mathematical model. It could also be a good article for students performing the optional statistical analysis described in Section 5. As part of a summative exercise, students could read it after they have performed the statistical analysis, to review why the concentration of swine facilities in poor and minority communities is detrimental to those communities.

Air-quality monitoring data specifically related to agricultural production is made publicly available by the EPA (though the limitations of these data have been criticized by GAO among others). The data for the state of North Carolina can be found here [15].

Additional resources for an extended unit analysis. For instructors who wish to develop an entire unit around this issue, the GAO report "Concentrated Animal Feeding Operations: EPA Needs More Information and a Clearly Defined Strategy to Protect Air and Water Quality from Pollutants of Concern" [6] provides detailed information regarding concentrated animal feeding operations in the United States, negative externalities associated with CAFOs including both air and water pollution, and the regulatory and legal challenges associated with addressing these externalities. The report totals 85 pages, but it is well organized and instructors could easily select specific sections to assign.

Preparation for environmental and social justice class meeting. This module is suitable in a differential equations or mathematical modeling course, after the students have studied systems of ordinary differential equations, and more specifically after they have seen interacting species. We recommend assigning the *Rolling Stone* article [14] prior to the class meeting, and preferably a more recent newspaper article on the relationship between farming practices and harmful algal blooms. We have used the Guardian article [5] on the water crisis in Toledo, OH, due to toxic algal blooms in Lake Erie.

4 The Module

One of the adverse effects of excess nutrient contamination attributable to industrial hog production is harmful algal overgrowth in affected waterways. Here, we present an in-class activity on mathematical models for the algae population dynamics, depending on the total rate of discharge of nutrients into the water, which affects dissolved oxygen and hence demonstrates the link between the pollution generated by hog farms and the death of marine life in coastal areas [12]. We then propose a class discussion on a more complicated model and suggestions for further investigations and projects.

There are many different scales at which this module could be implemented. An instructor might have students engage in light, topical reading prior to the single class period in which students engage with the mathematical or statistical problem motivated by the environmental justice issue highlighted in the reading. This approach has been successfully implemented in a 110-minute class. By assigning the reading in advance of the class meeting only a brief introduction to the topic is necessary prior to mathematical engagement. At the other end of the spectrum, an instructor could use the resources provided to develop a multi-week project in which students would engage seriously with the literature and develop a deep understanding of both the issues involved and the analytical techniques appropriate to analyzing the issue rigorously. The former approach may be better suited to a differential equations class, while the latter may be consistent with the aims of a modeling course.

4.1 The class. In the beginning, a class discussion on the assigned readings would be beneficial as a warm-up for students and to help them get started on thinking about the issues. After the initial warm-up, students can begin working on the *Dynamics of Algal Bloom Worksheet*. This handout is available in Appendix A.1 and solutions are provided in Appendix A.2. We have found that having students work in groups of three works best.

4.2 Post-worksheet discussion. The parameter values for Case 2, estimated from [12] are as follows:

$$\alpha_1 = 7, \ \alpha_2 = 0.1, \ \alpha_3 = 0.05, \ \alpha_4 = 2, \ \alpha_5 = 1.5, \ \beta_1 = .2, \ \beta_2 = 2, \ q = 8, \ q_C = 10.$$

A valuable exercise is to experiment with different initial conditions $(n(t_0), a(t_0), C(t_0))$ and lead a discussion on the effects and expectations of the students about the dynamics of the system. We include two figures below, demonstrating the behavior of the system with different initial conditions. The Matlab files used to create the figures are available online.

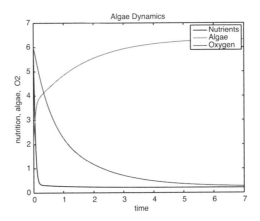

Figure 17.1. Sample figure for Case 1.

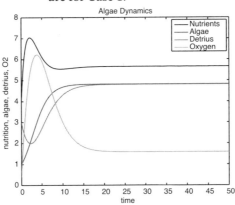

Figure 17.2. Sample figure for Case 3.

After working out the two models, groups can share their models and explain how they thought about the process. To sum up the class, the instructor could introduce the model that incorporates the detritus (waste) in addition to other dependent variables, as given in [**12**]. Due to the sophistication of the model in [**12**], we believe it best if the instructor walks the students through it, discussing why the model is set up the way it is. Below is the outline we have used.

4.3 Extensions. Mathematical modeling using differential equations most often yields non-linear equations or systems whose analytic solutions are not possible. There-fore, mathematical modeling of real world phenomena is a great opportunity to also discuss use of Matlab or other mathematical software. We recommend providing students a Matlab tutorial—there are many available for free online. We typically use one that includes the built-in `ode45` function which uses 4th order Runge-Kutta numerical scheme to solve ordinary differential equations. To reinforce students' learning programming with Matlab, an assignment that we like, following the module, asks the following questions:

(1) Compute the numerical solution for the two differential equations systems you found in Cases 1 and 2, using the initial conditions $n(0) = 1, a(0) = 3$ and $n(0) = 2, a(0) = 2, C(0) = 3$. Experiment with three different initial conditions.

(2) Plot the solution curves for $n(t), a(t)$ and $n(t), a(t), C(t)$ on the same plane, respectively.

(3) Plot n vs. a for Case 1, and n vs. C, and a vs. C for Case 2.

A more open-ended question is to ask students what other modeling techniques could be used for the situation they have seen and how they would improve the models they designed or were presented. The point of the open-ended or less-structured questions is to allow students the space to be creative and experience being mathematicians.

5 Additional Thoughts

Developing student proficiencies in data handling and analysis is a priority for many educators. Many scholars have argued that development of these proficiencies requires more than one or two dedicated courses [4]. Just as many schools now call for "writing across the curriculum", data handling and analysis must be more routinely incorporated into a wide range of courses. To this end, in the following, we describe a data handling and analysis module focused on the same issue studied earlier. This module enables instructors to provide students with an opportunity to conduct analyses on a meaningful database of significant size.

This is an extremely flexible module, and thus the prerequisites required range from virtually nil, to a 200-level statistics course, to a substantial programming (syntax writing) experience. The major contribution is the development of a "ready-to-analyze" dataset which demonstrates the disparate impact of the negative externalities associated with industrial swine production on minority and low-income communities.

5.1 Data handling & analysis module: Instructor preparation. In addition to the background resources described in Section 3 above, the instructor may also wish to read "Environmental Injustice in North Carolina's Hog Industry" by Wing et al. [17], an EHP "article" (see above) which reports on a study utilizing permitted swine production facility data from North Carolina in 1996. The data used in this study were obtained from the same source as the facility data incorporated in the database developed for the statistical component of this teaching module . Our data are more recent, however, reflecting permitted facilities as of January 2014. The analysis reported on in this article was conducted at the more granular block level. Instructors who employ the statistical component of the module may choose to have students read this article after conducting the statistical analysis. In this way students will "discover" the disparate impact on poor and minority communities through their own exploration of the data and could then read a formal presentation of a rigorous statistical analysis that controls for additional factors. Instructors using this module as the basis for a major unit of their course could choose to have students augment the database with additional variables motivated by [17].

Instructors will also benefit from familiarizing themselves with the dataset provided (described below) and possibly working through the scaffolded exercise prior to assigning it. It can be adapted as necessary to align with the prior experience of the targeted student population.

5.2 Data handling & analysis module: The dataset.

We have compiled a dataset which combines publicly available data on the location, head count, and associated manure lagoons of permitted swine production facilities in North Carolina, with Census-tract-level data on key demographic variables including race and poverty indicators from the American Community Survey (ACS). The permitted swine production facility data were obtained from the Animal Feeding Operations Branch of the North Carolina Division of Water Resources; see `http://deq.nc.gov/about/divisions` `/water-resources/water-resources-permits/wastewater-branch/animal-` `feeding-operation-permits/permits`. This is a micro-level data set, with an entry for each permitted operation activity (a given farm may hold multiple permits for different operation activities, such as "Farrow to Wean" or "Feeder to Finish", each of which has its own allowable head count); each entry is geocoded with longitude and latitude coordinates. TAMU geoservices, from Texas A&M, matched the latitude and longitude data to the associated census tract so that hog farm data could be combined with tract-level demographic and income data from the ACS. We produce multiple datasets for use with this module, and suggest a number of different activities depending on the course time available and the skill level of the students (or course goals).

The final database available for instructor use with this module is a tract-level database, including key demographic variables from the ACS and summary counts of the number of permitted operation activities, swine head count, and number of lagoons per tract. This database is ready for analysis and as such requires little data-handling from students. This dataset is ideal for engaging students with both basic descriptive data inquiries and simple analytical procedures such as chi-squared analysis and regression analysis. The database as provided includes calculated indicator variables such as "MINMAJ", which stands for "Minority Majority", indicating a tract in which the minority population is more than half of the total population. Instructors could leave such variables in the database if the activity goal was limited to statistical analysis. Alternatively, instructors could remove these variables from the database and require students to identify and then generate key variables required for analysis.

A pared-down database, limited to key variables, is provided for use with the simple linear regression module using Excel. Instructors who are able to have their students engage with the data using a statistical package may choose to use either the restricted database, `Swine_basic.xlsx` (which includes a data dictionary on a separate tab) or the more detailed database, `Swine_complete`, which is provided in both Excel and SPSS formats. The SPSS database includes variable labels. These files are available for download from

`https://drive.google.com/open?id=0BzUJgTOKk_uAdFcOSXcxTmpDSmM`.

The example module provided below motivates student interest in the topic by assigning advance reading of [14], a *Rolling Stone* article on the toxic effects of industrial swine production. The in-class handout assumes that the instructor will introduce (or refresh) student understanding of Ordinary Least Squares regression. Because different campuses use different software packages, the example module references Excel. Requiring students to import the provided Excel-format data into the statistical package commonly used in their own program, and to write syntax to perform various analyses, would raise the level of the required work.

5.3 Preparing students for the class meeting. We will refer to the class session in which students engage with the data to discover the disparate impact of industrial agriculture on low-income and minority communities as the "lab class". We recommend assigning the *Rolling Stone* article [**14**] prior to the lab class, and also asking students to review OLS prior to the lab. If students will be using a statistical software package (rather than Excel) we recommend informing them about which package they will be using and pointing them to resources for that package in case they are not familiar with it. For example, some instructors might choose to use R, which is available for free and is becoming more widely used across a variety of disciplines. In that case, instructors could encourage students to visit German Rodriguez's site "Introducing R" (available at `http://data.princeton.edu/R/`) which provides a solid introduction to the package.

Sample language to post on the course learning management system (or to include on a handout) in advance of the lab is provided below.

> Prior to our class session investigating environmental and social justice issues, please read the *Rolling Stone* article "Boss Hog: The Dark Side of America's Top Pork Producer". This 2006 article discusses the increasing concentration of livestock production in the United States and the negative environmental impacts that result from industrial agriculture, with a specific focus on pork production in North Carolina. You can access this article at `http://www.rollingstone.com/culture/news/boss-hog-the-dark-side-of-americas-top-pork-producer-20061214`.

> You should also refresh your understanding of basic linear regression and the ordinary least squares (OLS) method in particular. Numerous resources are available online, including [insert your preferred online resource(s) here]. There are a number of reasons one might engage in regression analysis, including describing a quantified relationship between observed factors (i.e., modeling), testing a hypothesis about how two variables of interest are related, and making predictions about future outcomes (out-of-sample predictions). Our focus in the next class session will be on using OLS to test a hypothesis.

5.4 The lab class. The *Rolling Stone* article [**14**] makes the negative impacts of industrial swine production clear, without explicitly pointing to the environmental justice issue. This provides a great opportunity for students to work together in groups to identify what factors would lead us to conceive of this as an environmental (or social) justice issue, rather than simply as an externalities problem. We provide two questions below to guide student thinking on this issue and to prepare them to work with the data. Though instructors could assign the following as either individual or group work, we generally prefer the latter as students can engage more deeply with the ideas and problem solve collaboratively. We suggest distributing the first two questions prior to making the data and statistical exercises available.

(1) The *Rolling Stone* article you read prior to class describes negative environmental externalities associated with industrial livestock production—in particular the

case of intensive hog operations in North Carolina. Why might we conceive of this as an environmental justice (or social justice) problem rather than a strictly environmental problem?

(2) What type of data would you need to determine whether the negative externalities associated with industrial hog production in North Carolina are appropriately characterized as an environmental justice issue?

The example data exercise we provide relies upon Excel, as this software is widely available, familiar to most students, and facility with Excel is a transferable skill. That said, we believe that students would benefit from more experience working with statistical packages. As noted above, if there is no institutional incentive to use a specific statistical package, R is available at no cost and is becoming more widely used across disciplines, especially in the sciences. The detailed regression exercise, available online as a modifiable Word document (`https://drive.google.com/open?id=0BzUJgTOKk_uAdFcOSXcxTmpDSmM`), provides extensive instruction and scaffolding which includes screenshots of Excel, as well as prompts to which students must respond. The amount of scaffolding provided is reduced as the exercise advances, so that students can apply on their own that which they have already learned.

5.5 After the lab. As a follow-up to the lab class, instructors may want to assign the article "Environmental Injustice in North Carolina's Hog Industry" by Wing et al. [17], which reports on the findings of a similar, though more detailed analysis using somewhat older data. Instructors could design a writing assignment integrating this article with the students' own analyses, or use the reading as part of an integrative discussion or group-work assignment in the following class period.

5.6 Extensions. The data set produced by NCDWR is a micro-level data set, with an entry for each permitted operation activity. We have made available the "enriched" micro-database, `Swine_micro`, which includes census-tract coding, for instructors who want to engage students in higher-level data handling operations. Upper division students with significant experience in database construction could be challenged to transform the database into a tract-level database to merge with the ACS demographic and economic data. The instructor can extract the ACS data from the `Swine_complete` file.

Alternatively, instructors may want to take advantage of this interesting micro-level data set to engage students with geographic information systems (GIS) analysis. Expertise in GIS is an increasingly important skill for graduates in many areas, including policy, resource management, ecology, and business. Students could use GIS to map the location of the permitted hog facilities in North Carolina to see where these facilities are concentrated. They could then read the 2000 article "Environmental Injustice in North Carolina's Hog Industry" by Wing et al. [17], which includes such a map for 1998 permitted hog facilities. This dataset is named "`swine_micro`" and is available in both SPSS and Excel format. All referenced files are available at the link provided in the Instructor Preparation section above.

Appendix A: Assignments and Handouts

A.1 The dynamics of algal bloom worksheet. NOTES TO INSTRUCTORS ARE IN BRACKETS, WRITTEN IN ITALIC FONT. AN ELECTRONIC VERSION OF THE WORK-SHEET IS AVAILABLE ONLINE AT

```
https://drive.google.com/open?id=0BzUJgTOKk_uAdFcOSXcxTmpDSmM.
```

The North Carolina hog industry is estimated to process 9-10 million animals per year [**10**]. The lagoons where the hog waste is collected house many nutrients, such as nitrates, and phosphate, in addition to antibiotics and other pollutants. In North Carolina's recent history, during three hurricanes, the manure lagoons flooded, and raw sewage spilled in the waterways [**8**]. The excess nutrients contained in the sewage caused a rapid increase in the algal population, called *algal bloom*. As the algae die, the breakdown to nutrient components feed the bacteria that decompose it, and with more food available, the bacteria thrive and increase in number, which in turn depletes the dissolved oxygen in the water further. The algae cover the water surface, allowing for very little transfer of oxygen from the air into the water through diffusion, and even then, the oxygen goes back into the atmosphere before dissolving in the water. Lack of dissolved oxygen makes the water uninhabitable by marine life, causing fish and other organisms to die, in millions.

Scenario 1: Case of unlimited supply of nutrients.

(1) Making the following assumptions, write a system of differential equations that describe the relationship between the algae population and the dissolved oxygen in the water.

Assumptions

(a) There is abundant supply of nutrients in the water.

(b) Algae population density $a(t)$ increases at a rate proportional to its population $a(t)$, due to unlimited resources.

(c) Algae die naturally at a rate proportional to its population $a(t)$.

(d) Algae population decreases due to overcrowding proportional to the square of its population, $a(t)^2$.

(e) The dead algae turn into nutrients, utilizing (depleting) dissolved oxygen $C(t)$; therefore, $a(t)$ and $C(t)$ are inversely proportional.

(f) The rate of increase of dissolved oxygen in water is only through contact with air by the water surface, and photosynthesis by the plants, is constant, given by q_C.

(g) The concentration of C decreases from natural causes at a rate proportional to its concentration.

(2) Find the equilibria of the system you wrote in Question 1.

(3) Investigate the stability of this system.

Scenario 2: Case of limited nutrients.

(4) In this case, there is a constant flow of nutrients into the water, at a rate of q gallons/unit time, and its concentration at any given time is $n(t)$. The algae population density $a(t)$ is proportional to and wholly dependent on the nutrient concentration. We keep the rest of the dynamics unchanged from Scenario 1. Create a model that describes change in $n(t)$, $a(t)$, and $C(t)$. [*The non-linear nutrition-algae interaction term na is worth discussing in class.*]

Scenario 3: Incorporating nutrient concentration, algae population, the detritus, and the dissolved oxygen in the water.

(5) In addition to the assumptions given earlier in the worksheet, consider the following and create a new model:

(a) The detrius $S(t)$ increases the nutrition concentration, and depletes the dissolved oxygen concentration in the water.

(b) The detrius decreases at a rate proportional to its concentration.

A.2 Solutions to the worksheet for the instructor.

(1) Under these assumptions, a model should have the following general form:

$$\frac{da}{dt} = \beta_1 a - \alpha_1 a - \alpha_2 a^2 = ra - \alpha_2 a^2,$$

$$\frac{dC}{dt} = q_C - \alpha_3 C - \alpha_4 a,$$

where, α_i, β_i are constants of proportionality, and are both positive, and $r = \beta_1 - \alpha_1$ can be either positive or negative.

(2) The equilibrium solutions are where $\frac{da}{dt} = \frac{dC}{dt} = 0$, and solving these simultaneously gives us $a^* = 0$, $C^* = \frac{q_C}{\alpha_3}$ and $a^* = \frac{r}{\alpha_2}$, $C^* = \frac{q_C \alpha_2 - \alpha_4 r}{\alpha_3 \alpha_2}$.

(3) The stability of equilibrium solutions for non-linear systems are determined by the eigenvalues of the Jacobian matrix, evaluated at each equilibrium point. The first equilibrium $(0, \frac{q_C}{\alpha_3})$ is a sink if $r < 0$, i.e., $\beta_1 < \alpha_1$, and a saddle if $r > 0$, i.e., $\beta_1 > \alpha_1$, and the second equilibrium $(\frac{r}{\alpha_2}, \frac{q_C \alpha_2 - \alpha_4 r}{\alpha_3 \alpha_2})$ is a sink if $r > 0$, i.e., $\beta_1 > \alpha_1$, and a saddle if $r < 0$, i.e., $\beta_1 < \alpha_1$. For detailed calculations, please see Appendix A.3.

(4) Here is a system that will work:

$$\frac{dn}{dt} = q - \alpha_1 na + \beta_1 a,$$

$$\frac{da}{dt} = \beta_2 na - \alpha_2 a - \alpha_3 a^2,$$

$$\frac{dC}{dt} = q_C - \alpha_4 C - \alpha_5 a.$$

The parameters α_i indicate depletion, while β_i indicate an increase.

(5) Under the new assumption of Scenario 3, the model, now a system of four differential equations, takes the form:

$$\frac{dn}{dt} = q - \alpha_0 n - \frac{\beta_1 an}{\beta_{12} + \beta 11n} + \pi \delta S,$$

$$\frac{da}{dt} = \frac{\theta \beta_1 na}{\beta 12 + \beta 11n} - \alpha_1 a - \beta_{10} a^2,$$

$$\frac{dS}{dt} = \pi_1 \alpha_1 a + \pi_2 \beta_{10} a^2 - \delta S,$$

$$\frac{dC}{dt} = q_C - \alpha_2 C + \lambda_{11} a - \delta_1 S,$$

where the α_i's are depletion rate coefficients and $\beta_1, \theta_1, \delta$, and δ_1 are constants of proportionality and are positive. For a study of equilibrium solutions and stability analyses, we refer the instructor to the paper by Shukla et al. [12].

A.3 Investigating the stability of the system of ordinary differential equations in Scenario 1. The model for algal population - dissolved oxygen dynamics is as follows:

$$\frac{da}{dt} = \beta_1 a - \alpha_1 a - \alpha_2 a^2 = ra - \alpha_2 a^2, \tag{17.1}$$

$$\frac{dC}{dt} = q_C - \alpha_3 C - \alpha_4 a, \tag{17.2}$$

where, α_i, β_i are constants of proportionality, and are both positive, and $r = \beta_1 - \alpha_1$ can be either positive or negative.

The two equilibrium solutions are $a^* = 0, C^* = \frac{q_C}{\alpha_3}$ and $a^* = \frac{r}{\alpha_2}, C^* = \frac{q_C \alpha_2 - \alpha_4 r}{\alpha_3 \alpha_2}$.

For non-linear systems, the stability of equilibrium solutions is determined by the eigenvalues of the Jacobian matrix of the system (17.1). (More details can be found in [3].) The Jacobian matrix for the system (17.1) is:

$$J(a^*, C^*) = \begin{bmatrix} r - 2\alpha_2 a^* & 0 \\ -\alpha_4 & -\alpha_3 \end{bmatrix}.$$

The eigenvalues of the $J(a^*, C^*)$ are given by the roots of the characteristic polynomial:

$$\lambda^2 + (\alpha_3 - r + 2\alpha_2 a^*)\lambda + \alpha_3(-r + 2\alpha_2 a^*) = 0. \tag{17.3}$$

Case 1: $a^* = 0$.

Using the quadratic formula for solving the characteristic polynomial (17.3), we obtain the two eigenvalues;

$$\lambda_{1,2} = \frac{1}{2}\left((r - \alpha_3) \pm \sqrt{(\alpha_3 + r)^2}\right) = \frac{1}{2}(r - \alpha_3 \pm |\alpha_3 + r|). \tag{17.4}$$

The two eigenvalues are: $\lambda_1 = r, \lambda_2 = -\alpha_3$. Since $-\alpha_3 < 0$ by the initial assumption on the sign of depletion rates, the sign of r determines whether the equilibrium is a sink or a saddle. If $r < 0$, then the equilibrium is a sink, and if $r > 0$, then it is a saddle.

Case 2: $a^* = \frac{r}{\alpha_2}$.

With a similar calculation, the characteristic polynomial of the Jacobian can be written as

$$\lambda^2 + (\alpha_3 + r)\lambda + \alpha_3 r = 0.$$

The roots of the characteristic polynomials, which are the eigenvalues of the Jacobian at the equilibrium $(r/\alpha_2, (q_C \alpha_2 - \alpha_4 r)/(\alpha_3 \alpha_2))$ are:

$$\lambda_{1,2} = \frac{1}{2}\left(-(r+\alpha_3) \pm \sqrt{(\alpha_3 - r)^2} = -(r+\alpha_3) \pm |\alpha_3 - r|\right). \tag{17.5}$$

The two eigenvalues are $\lambda_1 = -r$, $\lambda_2 = -\alpha_3$. As in Case 1, sign of r determines the type of equilibrium. If $r > 0$, i.e., $\beta_1 > \alpha_1$, then the equilibrium is a sink, on the other hand $r < 0$, i.e., $\beta_1 < \alpha_1$ results in a saddle equilibrium.

Bibliography

[1] Bartlett, R.L. (1996) "Discovering Diversity in Introductory Economics," *Journal of Economic Perspectives*, **10**:2, pp. 141-153.

[2] Burkholder, J. *et al.* (2007) "Impacts of waste from concentrated animal feeding operations on water quality," *Environmental Health Perspectives*, **115**:2, pp. 308-312.

[3] Blanchard, P., Devaney, L., and Hall, G., *Differential Equations*, Cengage Learning; 4th ed., 2011.

[4] DeLoach, S. B. and Finn, J. E. (2012) "Creating Quality Undergraduate Research Programs in Economics: How, When, Where (and Why)," *The American Economics*, **57**:1, pp. 96-110.

[5] Goldenberg, S. (2014), "Farming Practices and Climate Change at Root of Toledo Water Pollution," *The Guardian*, August 3, 2014. Available via `http://www.theguardian.com/world/2014/aug/03/toledo-water-pollution-farm ing-practices-lake-erie-phosphorus`, accessed on July 26, 2016.

[6] Government Accountability Office. (2008) "Concentrated Animal Feeding Operations: EPA Needs More Information and a Clearly Defined Strategy to Protect Air and Water Quality from Pollutants of Concern," GAO Publication No. 08-944, Washington D.C.: US Government Printing Office. Available via `http://www.gao.gov/products/GAO-08-944`, accessed August 7, 2016.

[7] Herrero, Mario *et al.* (2009) "Livestock, livelihoods and the environment: understanding the trade-offs," *Current Opinion in Environmental Sustainability*, **1**:2, pp.111-120.

[8] Nicole, W. (2013),"CAFOs and Environmental Justice: The Case of North Carolina," *Environmental Health Perspectives*, **121**:6, pp. 182-189.

[9] Nixon, S. (1998), "Enriching the sea to death," *Scientific American*, **9**:3, pp. 48-53.

[10] North Carolina Department of Agriculture and Consumer Services, 2012 Agricultural Statistics, (2012).

[11] Osterberg, D. and David, W. (2004), "Addressing externalities from swine production to reduce public health and environmental impacts," *American Journal of Public Health*, **94**:10, pp. 1703-1708.

[12] J. B. Shukla, A. K. Misra, and Peeyush Chandra, *Modeling and analysis of the algal bloom in a lake caused by discharge of nutrients*, Appl. Math. Comput. **196** (2008), no. 2, 782–790, DOI 10.1016/j.amc.2007.07.010. MR2388732

[13] Singh, S.D. and Olson, B. (2013), "Smithfield Receives U.S. Approval for Biggest Chinese Takeover," *Bloomberg L.P.*, September 6, 2013. Available via `http://www.bloomberg.com/news/articles/2013-09-06/smithfield-receives-u s-regulator-approval-for-shuanghui-deal`, accessed on July 26, 2016.

[14] Tietz, J. (2006), "Boss Hog: The Dark Side of America's Top Pork Producer," *Rolling Stone*, December 14, 2006. Available via `http://www.rollingstone.com/culture/news/ boss-hog-the-dark-side-of-americas-top-pork-producer-20061214`, accessed on July 26, 2016.

[15] U.S. Environmental Protection Agency, "Air Quality Systems Date: North Carolina." Available via `https://www.epa.gov/afos-air`, accessed on August 6, 2016.

[16] Wilson, S. *et al.* (2002) "Environmental injustice and the Mississippi hog industry," *Environmental Health Perspectives*, **110**:Suppl 2, pp. 195-201.

[17] Wing, S. *et al.* (2000) "Environmental injustice in North Carolina's hog industry," *Environmental Health Perspectives*, **108**:3, pp. 225-231.

[18] Ziegert, A. L. (2000) "The role of personality temperament and student learning in principles of economics: Further evidence," *The Journal of Economic Education*, **31**:4, pp. 307-322.

18

Student Loans: Fulfilling the American Dream or Surviving a Financial Nightmare?

Reem Jaafar

Abstract. In 2016, the total U.S. student debt stood at 1.2 trillion dollars— the second-highest level of consumer debt behind mortgages. The average debt for a 2015 graduate is $35,051 [2,9]. In this project, students explore the reasons that pushed the student loans debt to a new high, the repercussions of this debt on students' future plans, and the long-term effect of this debt on society in general. In the process, they read articles, gather and analyze data from given sources, and write a reflective essay. This multi-pronged approach engages students in applying their mathematical knowledge while understanding the nuances of the federal student loan system, including the large profits involved. **Keywords:** Student loan debt, tuition increase, simple and compound interests, polynomial functions and their graphs, limitations of polynomial models, essays, quantitative reasoning.

Here we describe a three-part module that can be used in most pre-Calculus, College Algebra, or Quantitative Reasoning courses in three consecutive weeks. Its purpose is to help students understand issues related to student loan debt and mathematically analyze different data and charts to interpret information about the subject. Students are expected to analyze different polynomial models, find the best fit, and understand the impact of interest rates through a case study. Finally, an essay is required to assess students' overall understanding of the social justice themes and the relevance of mathematics in that context.

The issue of student loans is near and dear to a majority of students: 66% of college seniors in the United States have an average of $26,600 in student loans [16]. In 2016, the total U.S. student debt stood at 1.2 trillion dollars—the second-highest level of consumer debt behind mortgages. The average debt for a 2015 graduate is $35,051 [2, 9]. Through this project, students gain experience in applying their mathematical knowledge to analyze and better understand a socio-political issue that has real implications for their own lives.

1 Mathematical Content

The goals of the project are two-fold. First, students are expected to graph data points and fit them to different polynomial models (linear, quadratic, quartic). Second, students should understand the effect of interest rates when taking out a loan. Instructors may devote half an hour of class time to discuss student loan debt and how it relates to students in the class. This discussion motivates students to read the articles, extract and graph the required data, and analyze the resulting model. Through the reading assignments, students are expected to understand thoroughly the civic issue, sharpen their quantitative reasoning skills, evaluate claims based on numbers, and, more generally, hone their critical thinking skills.

For this project, students should have prior exposure to modeling with linear functions, which makes it easy to introduce higher-order polynomial functions. They should also have prior exposure to interest rates. The project presents an opportunity to review the differences between simple and compound interests and how an increase in interest rate can impact student loans repayment and thus alter future plans.

In this project, students study the properties of polynomial functions and their limitations in application. Especially in the context of a pre-Calculus course we can expect our students to be proficient in linear and exponential models. However, while many students understand the main ideas of modeling, we as instructors often fail to communicate to them the limitations of each method when applied to a real-life situation. This project explicitly focuses on the limitations of polynomial modeling.

The project can be assigned when polynomial functions are introduced. The activities involved raise awareness about student debt and college education while providing the opportunity to learn more about polynomial functions and their properties in comparison to functions encountered in the textbook.

2 Context / Background

2.1 Underlying Social, Political, and Economic Context. The issue of student loans and student debt is relevant to students nationwide who are finding it more

difficult to pay for college. It is important for them to be aware of the political, economic, and social background of the student loan industry to be able to make informed decisions.

According to Baum, Ma, and Payea of the College Board [3], "[a] college education does not carry a guarantee of a good life or even of financial security. But the evidence is overwhelming that for most people, education beyond high school is a prerequisite for a secure lifestyle and significantly improves the probabilities of employment and a stable career with a positive earnings trajectory". In the report, the authors present a plethora of data that imply that college-educated adults are more likely to receive health insurance, are more likely to be active and engaged citizens, and are more likely to move up the socioeconomic ladder. To back up this last claim, the authors show that of the adults who grew up in the middle family income quintile, 31% of those with a college degree moved up to the top quintile between 2000 and 2008, compared with just 12% of those without a four-year college degree [3].

College education offers individuals an opportunity to lead a better and healthier life [3]. It can also lead to a more equitable society: in 2011, the poverty rate for bachelor's degree holders was 5%, compared to 14% for high school graduates [3]. According to the Bureau of Labor Statistics, the weekly earnings of a worker with a bachelor's degree are 70% higher than those of a worker with a high school diploma. In 2013, the unemployment rate for college graduates over 25 years old was 4% compared to 7.5% for high school graduates; in 2015 the same rates were 2.8% vs. 5.4% [3, 20].

A college education in the United States comes with a cost that has been consistently rising at a rate faster than inflation. Financial aid has not kept pace with these tuition hikes. As a result, students are forced to take out loans to pay for their education. Between 2005 and 2012, the number of student borrowers increased 66%, and the average student loan balance increased by 49%—from $16,651 to $24,803. This rise coincides with an increase in tuition at different types of institutions: private four-year, public four-year, and public two-year institutions [4, 5]. Once they graduate, students start paying back their loans, which reduces their take-home pay and their overall purchasing power.

Given today's job market and typical starting salaries, college graduates are struggling to fulfill their American Dream. Many end up postponing starting a business or owning a home. According to the American Student Assistance, 55% of survey responders indicated that student loan debt affected their decision to purchase a home, and according to the New York Federal Reserve, thirty-year-old adults are more likely to have home-secured debt if they do not have student loans [1].

The issue came to light in 2013 when Congress allowed the interest rates on student loans to double for some borrowers. More specifically, on July 1, 2013, the interest rates on subsidized government loans for students doubled to 6.8%. Chris Hicks, of the Debt-Free Future campaign for Jobs With Justice, argued that this turned the Department of Education into "a profit-making machine" and that the student debt program's purpose "isn't about helping students or borrowers—it's about making profits for the federal government" [15]. After intense debates, the *Bipartisan Student Loan Certainty Act* of 2013 was signed by President Obama on August 9, 2013, lowering the interest rates for new borrowers drastically (3.86% for undergraduate students in 2014; currently at 4.29%).

On May 6, 2014, Senator Elizabeth Warren introduced the *Bank on Students Emergency Loan Refinancing Act*. The aim of the bill was to lower the interest rates of borrowers who took out loans prior to July 2013. Some of these rates were near 7%; other older loans were at 9%. The bill offered the opportunity for those students to lower their interest rates to match those that the government offers to new borrowers. The bill failed to pass.

2.2 The Institutional Setting. My institution, LaGuardia Community College, serves a diverse and non-traditional student body that hails from 158 countries and speaks 129 different languages. Around 42% of students are over the age of twenty-two. For the entering class of 2014, 41% worked part-time or full-time, and 61% received financial aid [**14**]. The challenges of paying for a college education are compounded by the fact that many students have family responsibilities and depend on financial aid to be able to attend. Open-admission makes community colleges for some students the only bridge to a four-year college and a college degree. The majority of students attending LaGuardia come from families with incomes of less than $25,000 a year.

The College has invested in programs like the President's Society to provide students world skills and connections [**12**]. LaGuardia graduates transfer to four-year colleges at triple the national rate, and family income increases an average of 17% upon graduation. The majority of those students who graduate do not incur student loan debt. As such, LaGuardia can be regarded as a gateway for students to achieve their dreams without compromising their finances. The College also has a foundation that assists students in financial distress due to unpredictable circumstances [**13**].

Part of this module was used in a College Algebra class, which is a prerequisite for pre-Calculus. Depending on the size of the class, reading students' work may be more or less time-consuming. The syllabi of both College Algebra and pre-Calculus on our campus require at least one project a semester, so the module was incorporated for discussions during class time and an additional hour was used to give students feedback after they completed Part 1. In pre-Calculus, students are required to use a computer algebra system, which is an additional tool they can take advantage of when graphing and fitting data.

3 Instructor Preparation

3.1 Suggested Reference Materials. Student loans laws can change with time, so instructors need to stay up-to-date with current political events. The project can still be used by studying the impact of any new laws and how they change the status quo. Below is a list of resources suggested for student and instructor use. Students should be welcome to choose other reliable sources of information that provide information and data pertinent to the module as long as they provide a citation. The reading assignments may be updated yearly based on new reports, laws, and data.

Articles 1a [4] and 1b [18]: Both readings will help students understand the different types of student loans and their distribution. 1b [**18**] contains a glossary of terms.

Article 2 [7]: This is a tabulation of data describing the changes in tuition and fees and room and board charges over time from 1973-74 through 2013-14.

Article 3 [1]: This is a report titled *Life Delayed: The Impact of Student Debt on the Daily Lives of Young Americans* and provides good context for the project.

Article 4 [17]: This article, titled *The Student Loan Debt Crisis in 9 Charts*, presents and summarizes the data relevant to the crisis.

Article 5 [10]: This article on the Bipartisan Student Loan Certainty Act of 2013, titled *Student Loan Act Could Mean Higher Federal Profits*, describes the possible financial profits of the federal student loan program.

Article 6 [15]: This article, titled *Student Loan Borrowers' Costs To Jump As Education Department Reaps Huge Profit*, also emphasizes the profit the government could make from student loans.

Video [21]: This is the video recording of the speech given by Senator Elizabeth Warren in May 2014 when introducing the *Bank on Students Emergency Loan Refinancing Act*.

Additional Sources: Other websites and organizations provide relevant information and updated data: The Project on Student Debt [11], The Bureau of Labor Statistics [19], The College Board [6], and The Consumer Financial Protection Bureau [8].

3.2 Grading Issues. When I used this module in my courses, I assigned a grade to each question in Parts 1 and 2. The guidelines clearly stated that students must use complete sentences to answer questions. If the students did not, they earned half the credit assigned.

As mathematics instructors are often not trained to teach or evaluate writing, Part 3 can be more difficult to grade, but a checklist (or a rubric) can ease the grading on instructors and the burden on students who might feel overwhelmed with the information presented. I recommend returning the checklist along with the project to indicate the missing items.

Some of my students complained about having to work on a writing assignment in a mathematics project, so it may be important to spell out the expectations and the differences between an essay in an English class and an essay in a mathematics class.

4 The Module

Module Overview. The project consists of three parts spread over a period of three weeks. In the first part, students graph data representing relevant quantities such as tuition at different types of institutions, find the best fit and analyze different polynomial models, and state their domain and range in application. In the second part, students are asked to read about the *Bipartisan Student Loan Certainty Act* of 2013 and its implications for students. They are also expected to understand the role of government in regulating the student loan industry. Here, an optional segment about interest theory can be integrated. In the third part, the project focuses on the effect of interest rates and the impact of tuition increases on one's ability to attend college through a case-study. It ends with an essay to assess students' ability to apply mathematical knowledge in the context of the student loan industry. Interest rates and compounding effects are prerequisite ideas, but the instructor can briefly review those concepts. Sample copies of all assignment handouts can be found in the appendices.

Before giving the assignments to students, instructors can hand out the reading material to prepare them. Ideas in the articles can later be introduced in class and discussed among peers or with the professor. At that point, instructors can pass out the entire project and briefly explain the different parts, the general expectations, and the writing assignment at the end. If the pre-Calculus course offers a computer lab, students can plot the data during the computer hour.

Before going into the mathematical activities of the module, I recommend that instructors give students a historical perspective on the student loan industry to be able to understand its current status. Alternatively they may briefly introduce the social themes towards the end of class time and give students one week to complete the readings outlined in Section 3.1. They can also hand out the project at that point simply to allow students to see the questions that are directly related to the reading.

One week later, the computer lab hour (or the class hour) can be used to ask students specific questions about the articles and to initiate a discussion about the social themes. Instructors can start with questions that give relevant information about the issue: the size of the student loan industry, how it evolved during the past decade, the increase in tuition at public and private colleges, interest rates on student loans, and recent laws that were enacted or debated. Then they can ask students if they themselves have loans and if they know what their monthly payment will be upon graduation. Instructors can also ask students about their yearly tuition and introduce the data of Article 2 [7]. Then instructors can follow up with mathematical questions related to concepts such as the domain and range, the best fit, and the meaning of linear, quadratic, and quartic models. Students can submit their work for Part 1 of the module prior to the due date, as a proof of progress and for feedback. The project includes a final writing assignment.[1]

After collecting students' work for the first part, instructors can proceed to talk about the *Bipartisan Student Loan Certainty Act* of 2013 and the *Students Emergency Loan Refinancing Act* that failed to pass in 2014. The discussion can be started by assessing students' prior knowledge about interest rates and how they use simple and compound interest in general. Then the formulas that govern both types of interest may be derived. One may introduce an online calculator and how it is used to calculate the interest accrued on students' loans in two different scenarios in Part 2 of the project.

Finally, the project ends with a reflective essay that ties the different parts of the project together. At that point, students are expected to use the data, charts, and calculations to argue for a position and to make an informed decision. Students may be given clear guidelines about the expectations of the project. These guidelines can be converted into a rubric (or checklist) that reflect different weights instructors wish to give to different parts of the essay.

Part 1a. Why Is College Tuition over 100% Higher than What It Was Twenty Years Ago? Above I described how the social and the mathematical themes can be infused into the course. It is also helpful to ask students to compare their current tuition with what their peers paid twenty years ago. Based on the ensuing class discussions, students should graph the data presented in Article 2 [7]. They can use Excel or Maple or any other technological tool that can plot and fit data points. This part of the assignment is detailed in Appendix A.1.

---

[1]Alternatively, students can turn in work only once, after completing all parts of the project. In the following I will assume that that students turn in part of their work after the first part and then the rest at the very end.

After graphing the tuition during the indicated period of time, students will obtain a linear fit for private four-year colleges; a quadratic fit works for public four-year colleges, and a quartic fit works for public two-year colleges; see Figure 18.1. The three functions are increasing over time.

Through this activity, students understand the idea of fitting data more profoundly. Specifically, when the data are fit to a quartic function, students realize that they cannot use the function to make long-term predictions. Polynomial modeling may provide good fits within the range of data but will deteriorate rapidly outside the range of the data. This way, students understand the limitations of high-degree polynomial models. The models can be used to make future projections and then students can assess the reliability of these projections. Instructors may also discuss poor interpolatory properties, high oscillations between data points for high-degree polynomials, and the number of parameters associated with them. Finally, they may introduce the idea of infinite limits for polynomial functions.

Part 1b. Reading Questions. In this section, students are asked to research some facts and answer the questions posed using complete sentences. Students read recommended handouts, extract information and the necessary definitions, translate the meaning of data, and relate some of the information presented in different charts and articles. The section reinforces their quantitative reasoning skills. A list of questions that might be used to create a worksheet, a handout, or an assignment related to this section can be found in Appendix A.2.

In reference to the questions about Article 4 [**17**], students should visually determine functions that might fit each data set well – they can be linear for the consumer price index and exponential (or possibly quadratic) for tuition. Although it is easy for students to see that tuition has been increasing faster than the consumer price index, they should, however, argue based on numbers in the graphs or based on the function that fits the data. It is also important for students to understand the meaning of numbers. Follow-up questions in class can include:

- *Give two examples of a trillion dollar "industry".*

- *What is the GDP of the United States? Compare it with the GDP of Canada, and a European country of your choice.*

- *Find a company whose revenue is higher than the GDP of a developed country of your choice. Does this fact surprise you? Explain.*

- *When the real estate market collapsed in 2008, mortgage debt (estimated at a little over 14.7 trillion at the time) constituted a certain percentage of the GDP.*

  (1) *Estimate this percentage.*
  (2) *How does student loan debt today compare with the GDP of the U.S.?*
  (3) *How does the current student loan debt compare with the mortgage debt of 2008?*

These questions about Article 4 serve as a transition to Part 2, which focuses on interest rates for loans.

Part 2a. The Impact of Interest Rates and the Political Debate. In this part, students learn in great detail about a bill that increased the cost of borrowing on students – but increased government profits.

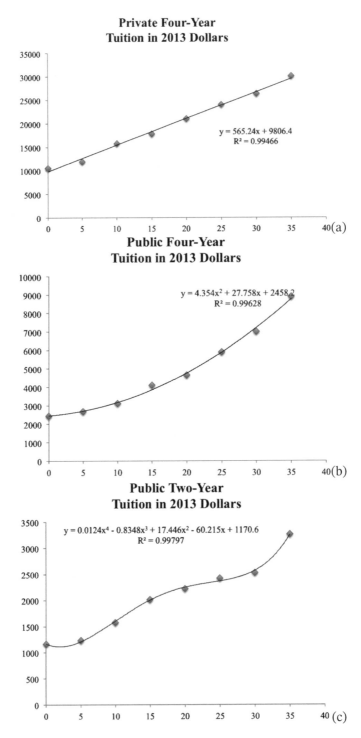

Figure 18.1. Tuition at different types of institutions between 1978-79 and 2013-14; figures created by the author for the data from Article 2 [**7**].

Students read articles [**10**,**15**], watch a video [**21**], explain bills debated in Congress, watch online videos, and extract information related to government profits from student loans. Students are expected to draw similarities between those profits and those generated by Fortune 500 companies. Projects should build students' critical thinking and digital literacy, and present an opportunity for instructors to give students a choice in exploring the answer to some problems. Appendix A.3 provides a list of questions that can be used in a handout or as prompts during class for this part of the assignment.

After watching the video [**21**], it is helpful to initiate a discussion on why people are able to refinance auto loans, home loans, and credit card loans at lower interest rates but not student loans.

The Bank on Students Emergency Loan Refinancing Act did not pass when a vote was held in Congress. This means that some of those who borrowed prior to July 1, 2013 will continue to pay high interest rates. This may be a natural tie-in to the topic of the effect of interest rates on loans, if this is a topic the instructor chooses to cover.

Part 2b. An Optional Discussion on Interest Theory. If the project is part of a College Algebra or a Quantitative Reasoning course, Appendix A.4 provides a generic list of questions that can be used to help students understand interest theory and use online calculators to compare payments for loans with different interest rates. This part is optional. The last question shows that a lower interest rate means long-term savings in interest owed. This idea is revisited below in Part 3. This section also exposes them to the concept of interest before they use the online calculator in Part 3.

Part 3a. The Impact of Attending College Fifteen Years Apart. In this part, students use an online calculator to determine the interest accrued and monthly payment. A list of questions that can be used to create a handout for this part is provided in Appendix A.5.

Students imagine a scenario involving two students who attended college fifteen years apart. They will be guided in filling out a table to assess the impact of student loans in two different cases. Each student needs to make a choice about the type of institution. The aim of this activity is not only to show students the impact of interest rates on a fixed amount, but also to show the impact of borrowing in different political and economic environments. The amount borrowed will be 50% of the tuition of each student. Instructors can modify the academic year or the type of loans (by changing the interest rates). Appendix A.5 details the assignment and shows a Sample Table.

After completing the table shown in Appendix A.5, students should conclude that although Jennifer has a lower interest rate than John, she will owe $9028 more than John due to the rising cost of education. As a result, she owes more in interest ($4,110.57 versus $4,870.17).

Part 3b. The Same Tuition, Two Different Interest Rates. In this section, students calculate the monthly payments and total interest paid for Jason, a classmate of John. A list of questions that can be used to create a handout for this part is provided in Appendix A.6.

To see the impact of doubling the interest rates, we assume Jason had a 3.46% interest rate and borrowed the same amount as John. A sample table is provided in Appendix A.6. Instructors can ask students about Jason's savings compared to John's and interpret the magnitude of the amount.

Part 3c. Student to Student: How to Avoid a Financial Nightmare. In this part, students need to write a letter of 500 to 800 words to Jennifer explaining her options. A prompt and some guidelines for this exercise can be found in Appendix A.7.

Depending on students' academic levels, instructors may or may not structure the guidelines. One way to scaffold this part is to require a three-part essay; this is how I structured the prompts in Appendix A.7. In Part 1, students can give an overview of the student loan industry and recent bills that passed. In Part 2, students can introduce Jennifer's case as outlined in the table (the type of institution she attended, her estimated monthly payment, etc.). In Part 3, they can explain to Jennifer her options.

Students should formulate a conclusion based on evidence they present.

5 Additional Thoughts

Here we list a few ideas on how this module can be extended.

The module suggests one venue to highlight student loan debt while using traditional mathematical concepts. It can be expanded or modified to account for other cases. Examples of such cases are: the cost, rewards (or lack thereof) of attending for-profit colleges, students' savings from attending a two-year college before transferring to a four-year college in New York, and a comparison of the student John's loan size with another student who attended a different type of institution at the same time.

One may also draw a parallel between credit card and student loan debt and compare the effect of each on one's personal finances.

In understanding the limitations of the polynomial models created, the instructor can introduce the idea of limits for students in pre-Calculus. An additional section can be appended if the course requires students to master a computer algebra system.

Appendix A: Assignments and Handouts

A.1 Assignment for Part 1a. The following questions can be used in a handout for Part 1a of this project. Refer to [7].

Why is college tuition over 100% higher than what it was twenty years ago?

(1) Graph the tuition and fees for private non-profit four-year colleges, public four-year and public two-year colleges, starting from the academic year 1978 -79 as year 0, and going in five-year intervals.

(2) Fit the data to a polynomial function (of least degree).

(3) What is the common trend for the three graphs?

(4) What is the domain and range of each function? For each one, give an example of a similar type of function seen in class (or in the textbook), and state the domain and range of each one.

(5) Use this model to predict the tuition cost for a student entering each type of institution in the academic year 2020-2021.

(6) Pick and explain how the data from Article 2 would validate the title of Part 1a: *Why is college tuition over 100% higher than what it was twenty years ago?*

(7) List the properties of polynomial functions, and explain the advantages for using them in modeling.

(8) From Figure 18.1, explain the limitations of polynomial fitting. Specify at least three limitations and consider the values of the function as x increases.

(9) When does polynomial modeling work best?

A.2 Assignment for Part 1b. The following questions can be used in a handout for Part 1b of this project. For questions 1 to 4, refer to articles [**4**, **18**]; for questions 5 to 7 refer to [**1**]; and for questions 8 to 13, refer to [**17**]. The numbering of articles aligns with that provided in Section 3.1.

Reading Questions

(1) Explain the difference between subsidized and unsubsidized federal loans.

(2) When did the federal government begin giving unsubsidized loans to students according to the chart in *Trends in Student Aid 2015*?

(3) Between the years 2000 and 2012, what was the trend for both subsidized and unsubsidized loans?

(4) Define private student loans.

(5) State five ideas that were surprising or interesting to you when reading the article and explain in what way(s) you found them interesting/surprising. How do these ideas relate to Figure 18.1?

(6) Towards the end of the article, the author lists some recommendations for actions. Summarize these recommendations and explain what can be done at the federal, state and private sector level in order to fund higher education.

(7) Which of the recommendations do you agree with the most? Choose one and explain your position using quantitative arguments.

(8) Choose five charts in the article and write a 200-to-300-word paragraph explaining the meaning of the data presented in the graphs and how the information presented in each are related to one another.

(9) Explain how the information in Article 4 reinforces the data of Article 2.

(10) In the figure that represents the percentage change of tuition and The Consumer Price Index (CPI), explain the two trends and suggest what type of functions can be used to fit each curve. Which trend is rising faster? Give a mathematical argument.

(11) What is the size of the student loan debt? How does it compare with the auto loan and the credit card industry?

(12) Present the data of "who owes the most" in a pie chart.

(13) Compare the different scenarios for the total interest paid on a $23,000 loan.

A.3 Assignment for Part 2a. The following questions can be used in a handout for Part 2a of this project. Refer to articles [10, 15] and the video [21].

The Impact of Interest Rates and the Political Debate

(1) From Article [10] (or any other source of information of your choice), explain what the Bipartisan Student Loan Certainty Act of 2013 means for students.

(2) From the video posted on the link of Article [15]: How much is the Department of Education profiting from student loans? In a brief paragraph, paraphrase how the reporter explains how the DOE will profit.

(3) According to Article [15], the "U.S. Department of Education is forecast to generate $127 billion in profit over the next decade from lending to college students and their families, according to the Congressional Budget Office". How much would those profits be per year? Compare the expected yearly profit of the student loan industry to the yearly profit of a Fortune 500 company of your choice. For reference, choose a recent year.

(4) What is the interest rate for an undergraduate borrower?

(5) According to Senator Warren's video [21], how many borrowers have student loan debt? Does this number reinforce information from the articles you read?

(6) What is the Banking on Students Emergency Loan Refinance Act? Why is it important to lower the interest rate on student loans? Please explain.

(7) How would the bill have been funded if enacted?

(8) What is Senator Warren's ultimate goal? Why does she think that education is important?

A.4 Assignment Part 2b: Optional. The following questions can be used in a second handout for Part 2 if the instructor wishes to include an optional section on interest theory.

An Optional Discussion on Interest Theory

Suppose you need to borrow money from your classmate. Your classmate agrees to lend you $500, but you will accrue 1% in interest every day.

(1) Write an expression for the amount you owe after 10 days, and generalize it for t days as a function of t.

(2) How much did you owe in interest alone? What is the name of this type of interest?

(3) What is the difference between simple and compound interest? How is the compound interest calculated?

(4) Consult the link below to learn how the interest on student loans is calculated:

$$https://studentaid.ed.gov/sa/types/loans/interest-rates\#what-interest.$$

Next, calculate and compare the interest owed on the following two loans:

Loan A: loan with principal amount of $15000 with interest at 7%, assuming you made your last payment 30 days ago.

Loan B: loan with principal amount of $15000 with interest at 4%, assuming you made your last payment 30 days ago.

A.5 Assignment Part 3a: The Case of John and Jennifer.

The following can be used in a handout for Part 3a of this project. The link to the online calculator is: `http://www.finaid.org/calculators/loanpayments.phtml`.

The Impact of Attending College Fifteen Years Apart: The Case of John and Jennifer

Imagine the following scenario. John attended college between 1999-2003. Jennifer is planning to enroll in the fall of the current year and expects to graduate four years later. Assume that the amount borrowed constitutes 50% of the tuition.

(1) Fill in Table 18.1.

Table 18.1. The case of John and Jennifer.

John and Jennifer			
	John	Jennifer	Note
Tuition			You need to find this information.
Loan Amount			John and Jennifer borrowed the same percentage (50%) of tuition.
Loan Term	10 years	10 years	
Interest Rate	6.92%	4.66 %	
Monthly Payment			Use the online calculator.
Total Interest Paid			Use the online calculator.
Cumulative Payments			Use the online calculator.

(2) In referring to John and Jennifer, who pays less in interest and by how much?

(3) After repaying their loans, what are the cumulative payments of John and Jennifer?

(4) In reference to the previous question, is this result in line with or contradictory to the interest rate each borrower got?

Here is what a sample table might look like when filled in:

Table 18.2. Sample case study of John and Jennifer if they both attended a four-year public college, fifteen years apart.

The case study of John and Jennifer if they both attended a four-year public college, fifteen years apart.			
	John	Jennifer	Note
Tuition	$5292	$9637	Both Attended public four-year colleges. For John: Using $f(x) = 4.354x^2 + 27.758x + 2458.2$ and calculating f(21), f(22), f(23) and f(24), one finds that the average tuition was $5292 rounded to the nearest dollar. For Jennifer, calculating f(36) through f(39) yield an average tuition of $9627.
Loan Amount	$10,584	$19,254	$5292 * 4/2 = $10,584$ and $9627 * 4/2 = $19,254$
Loan Term	10 years	10 years	
Interest Rate	6.92%	4.66 %	
Monthly Payment	$122.45	$201.03	Use the online calculator.
Total Interest Paid	$4,110.57	$4,870.17	
Cumulative Payments	$14,694.57	$24,124.17	

A.6 Assignment Part 3b: The Case of John and Jason. The following can be used in a handout for Part 3b of this project. The link to the online calculator is: `http://www.finaid.org/calculators/loanpayments.phtml`.

Same Tuition Different Interest Rates:
The Case of John and Jason

Imagine the following scenario. John and Jason attended college concurrently. They borrow a comparable sum of money. However, Jason received a lower interest rate on his loan. Assume that the amount borrowed constitutes 50% of the tuition.

(1) Fill in Table 18.3.

Table 18.3. Table to be filled for this part.

John and Jason			
	John	Jason	Note
Tuition			The same.
Loan Amount			Jason and John borrowed the same amount.
Loan Term	10 years	10 years	
Interest Rate	6.92%	3.46 %	
Monthly Payment			Use the online calculator.
Total Interest Paid			Use the online calculator.
Cumulative Payments			Use the online calculator.

(2) How much does Jason save in interest compared with John? Is the amount significant?

Here is what a sample table might look like when filled in:

Table 18.4. Sample case of John and Jason, two classmates with two different interest rates.

John and Jason: Two classmates with two different interest rates		
	John	Jason
Tuition	$5292	$5292
Loan Amount	$10,584	$10,584
Loan Term	10 years	10 years
Interest Rate	6.92%	3.46 %
Monthly Payment	$122.45	$97.93
Total Interest Paid	$4,110.57	$1,829.42
Cumulative Payments	$14,694.57	$11,751.42

A.7 Assignment Part 3c: Essay Prompt and Guidelines. The following is a prompt and a set of guidelines I use for the essay in Part 3c of this project.

Essay Prompt and Guidelines:
How to Avoid a Financial Nightmare

According to Senator Elizabeth Warren, student loans are preventing graduates from "making the purchases that keep this economy moving forward" [21]. In Article 6 [15], Chris Hicks claims that "The student loan program isn't about helping students or borrowers – it's about making profits for the federal government".

While being a student at a Community College, you might have taken loans or be planning to upon transferring, and you did this project to become an informed citizen.

Jennifer, who is college-bound next fall, is uncertain about her future and long-term goals. She contacts you for advice. She would like to understand the student loan industry in general, college tuition, and how political and economic factors might impact her long-term goal of buying a home.

General Guidelines: Write a coherent essay; your arguments must be based on numbers and data. You are required to use at least 10 different numbers in context and explain at least four different trends (using charts) in context. You are also required to use two different quotations.

(1) **In Part 1:** *Explain the recent trends you read about in the articles. Use at least 3 different charts to support your claims. Explain how the government will make profits from student loans and the crisis of 2013 before passing the Bipartisan Student Loan Certainty Act.*

(2) **In Part 2:** *Choose a major for Jennifer and research salary expectations for her major. Estimate her take-home pay. Estimate Jennifer's other expenses and estimate her savings per year.*

(3) **In Part 3:** *Estimate whether Jennifer will be able to save 20% of the cost of a home to pay the down payment, ten years after graduating. Choose a geographical area for Jennifer to live and research home prices in that area. Support your advice with data.*

(4) **In conclusion:** *What is your advice for Jennifer? Do you think student loans are preventing your generation from buying homes or from starting businesses?*

Bibliography

[1] American Student Assistance, *Life Delayed: The Impact of Student Debt on the daily Lives of Young Americans*, report, January 1, 2015, available at `http://www.asa.org/for-partners/schools/content-pages/life-delated-the-i mpact-of-student-debt-on-the-daily-lives-of-young-americans/` (requires registration), accessed on April 18, 2016. The file can also be retrieved from `http://www.asa.org/site/assets/ files/3793/life_delayed.pdf`, last accessed on October 26, 2016.

[2] Barr, C., *Student Loan Resources: Financial Aid & Loan Debt Management for Students*, web content, available at `https://www.debt.org/students/`, accessed on September 14, 2016.

[3] Baum, S., Ma, J., and Payea, K., *Education Pays 2013: The Benefits of Higher Education for Individuals and Society*, The College Board Report, January 2013, available at `https://trends. collegeboard.org/sites/default/files/education-pays-2013-full-report.pdf`, accessed on September 14, 2016.

[4] Baum, S., Ma, J., Pender, M., and Bell, D., *Trends in Student Aid 2015*, The College Board Report, January 2015, available at `http://trends.collegeboard.org/sites/default/files/ trends-student-aid-web-final-508-2.pdf`, accessed on September 14, 2016.

[5] Baum, S. and Payea, K., *Trends in Student Aid 2013*, The College Board Report, January 2013, available at `http://trends.collegeboard.org/sites/default/files/student-aid-2013-full-report. pdf`, accessed on September 14, 2016.

[6] The College Board, *Trends in Higher Education*, web content, available at `https://trends. collegeboard.org/`, accessed on September 14, 2016.

[7] The College Board, *Tuition and Fees and Room and Board over Time, 1975-76 to 2015-16, Selected Years*, web content available at `https://trends.collegeboard.org/college-pricing/figures-tables/tuition-and-fee-and-room-and-board-charges-over-time-1973-74-through-2013-14-selected-years`, accessed on September 14, 2016.

[8] Consumer Financial Protection Bureau, *Students and young consumers*, web content, available at `http://www.consumerfinance.gov/students/`, accessed on September 14, 2016.

[9] Edvisors, *What is the average amount of a student loan?*, web content, available at `https://www.edvisors.com/ask/faq/average-amount-student-loan/`, accessed on September 14, 2016.

[10] Equal Justice Works, "Student Loan Act Could Mean Higher Federal Profits", *U.S. News & World Report*, web content, August 14, 2013, available at `http://www.usnews.com/education/blogs/student-loan-ranger/2013/08/14/student-loan-act-could-mean-higher-federal-profits`, accessed on September 14, 2016.

[11] The Institute For College Access and Success, *Project on Student Debt*, web content, available at `http://projectonstudentdebt.org/`, accessed on September 14, 2016.

[12] LaGuardia Community College, *Meet Our Remarkable Students*, web content available at `http://www.laguardia.edu/About/Our-Students/`, accessed on September 14, 2016.

[13] LaGuardia Community College, *Foundation Scholarships*, web content available at `http://www.laguardia.edu/scholarships/`, accessed on September 14, 2016.

[14] LaGuardia Community College Office of Institutional Research & Assessment, *LaGuardia Community College Institutional Profile 2015*, available at `http://www.lagcc.cuny.edu/IR/IR-facts/`, accessed on September 14, 2016.

[15] Nasiripour, S., "Student Loan Borrowers' Costs To Jump As Education Department Reaps Huge Profit," *The Huffington Post*, April 14, 2014, available at `http://www.huffingtonpost.com/2014/04/14/student-loan-profits_n_5149653.html`, accessed on September 14, 2016.

[16] Reed, M., and Cochrane, D., *Student debt and the class of 2011*, The Institute for College Access & Success report, available at `http://ticas.org/sites/default/files/pub_files/classof2011.pdf`, accessed on September 14, 2016.

[17] Severns, M., "The Student Loan Debt Crisis in 9 Charts", *Mother Jones*, web content, June 5, 2013, available at `http://www.motherjones.com/politics/2013/06/student-loan-debt-charts`, accessed on September 14, 2016.

[18] SimpleTuition by LendingTree, *Student Loan and Consolidation Glossary*, web resource, available at `http://www.simpletuition.com/glossary`, accessed on September 14, 2016.

[19] United States Bureau of Labor Statistics, website, available at `http://www.bls.gov/`, accessed on September 14, 2016.

[20] United States Bureau of Labor Statistics, *Employment Projections: Earnings and unemployment rates by educational attainment, 2015*, web content last updated on March 2016, available at `http://www.bls.gov/emp/ep_chart_001.htm`, accessed on September 14, 2016.

[21] Warren, E. *Student Loan Refinancing Floor Speech*, video, May 6, 2014, available at `https://www.youtube.com/watch?v=aC6N-bU2jiM`, accessed on September 14, 2016.

19

Modeling Social Change: The Rise in Acceptance of Same-Sex Relationships

Angela Vierling-Claassen

Abstract. According to surveys by Gallup and the Pew Research Council, acceptance of same-sex relationships has risen dramatically over the past ten years. Mathematical tools can be used to study how this kind of societal change comes about. This paper describes a classroom activity that can be used to model the spread of acceptance of same-sex relationships, by representing a social network with a random graph and using a dynamic process to alter opinions and rewire social connections. This model can then be assessed by students for how well it predicts current social changes, and the parameters of the model can be altered to determine how changes in the attributes of people (nodes) and edges (social ties) impact the acceptance level and community over time. A follow-up exploration of a more robust computer model is included. Such a project can involve students at a wide variety of skill levels, creating opportunities for discussion and cooperation between students in both introductory and advanced courses.

Keywords: Same-sex marriage, LGBT acceptance, graph theory, networks, role-playing, group discussion.

1 Mathematical Content

This classroom activity has students simulating a social network and a process of opinion change through network adjustment. Students can then observe the outcome of the process on a larger scale through a network model run in Octave/Matlab and the programming language R. No knowledge of R or Matlab is necessary for the students or the instructor, but altering the parameters in the code

(available at `https://liberationmath.wordpress.com/LGBTopinion/`)

to create additional images and models will require some minimal programming experience.

The activity described here requires little mathematical background, so it could be used in a variety of courses and classroom situations. I have used the module in courses on game theory and liberal arts mathematics; it has also been used as an activity for a meeting of an LGBT (Lesbian, Gay, Bisexual, and Transgender) student group. The activity would also work well as part of a graph theory or discrete mathematics course. For more advanced students interested in alterations to the programming of the computer model, familiarity with basic programming would be useful. For students wishing to explore more about the mathematics of social networks, see [5], [6], and [8]. For students interested in opinion formation in social networks, see [2].

Through the activity and interacting with the model, students can work with setting up graphs and basic mathematical ideas such as randomness, averages, and measures of central tendency. Exploring a random graph as a model for a social network, students can start to understand local versus global properties as well as self-organizing behavior.

2 Context / Background

According to a 2013 Gallup poll, the percentage of Americans who believe it is "morally acceptable" to be in a same-sex relationship rose from 40% to 58% between 2000 and 2013 [3]. A Pew Research poll had similar results, with the number of Americans who believe that "homosexuality should be supported by society" growing from 47% to 57% between 2003 and 2013 [7]. Support has also grown, much more slowly, for transgender people [4]. With such a large shift, it is natural to wonder how and why such a change has happened. While media attention and advocacy by LGBT (Lesbian, Gay, Bisexual, and Transgender) organizations have been part of the driving force, relationships have also been a key driving factor. When a person has a relationship with someone they know is lesbian, gay, bisexual, or transgender, it has an impact on their opinions. For more about the history of the LGBT movement in the United States, see [1].

3 Instructor Preparation

3.1 Before Class Meeting. Before class, the instructor will need to set up a random network with the number of nodes equal to the number of students in the class. This network should include information about the acceptance level and connections for each node (see Figure 19.1 for a small sample network and the appendix for a sample information sheet about a node). If there are absent students, a few capable students can work with two nodes. I started my first simulation with 10–12 connections per student, but that was too many, so I decreased that to an average of around six connections for each node. Some students will still have more than six, so using a 12-sided die during the activity is useful, and there should be one available for each student.

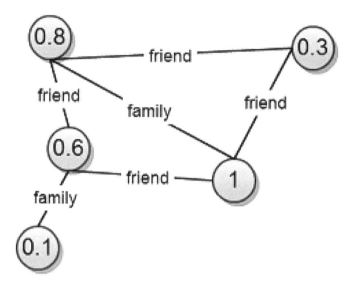

Figure 19.1. Example network with acceptance levels and edge types.

It is possible to create such a network by hand or using a spreadsheet program. To create the network in a spreadsheet, it makes sense to use two columns, Vertex 1 and Vertex 2. To create a random graph of N nodes in Excel, one can enter =int(rand()*N) into each cell in the two columns (a similar function should work in other spreadsheet programs). The number of rows divided by N will give the average number of ties for each node, so to get an average of six edges per node, one should create $6N$ rows. These ties also need to be coded as "friend" or "family" which can be done randomly by assigning each tie a random number between 0 and 1 using the function rand() and then having the family ties be the ones where this number is above (0.9). This will give about 10% family ties, and this is one of the many parameters that can be adjusted in the model. The nodes are all assigned numbers 1 to N, and one or two nodes should be randomly selected to be LGB nodes with an acceptance level of 1. The rest should be assigned a starting acceptance level using rand().

It is nice to have a visual for the network to show the class. If the network is created in Excel, NodeXL can be used to create the visual. If the network is created in Matlab/Octave, it can be visualized in R (code for this is available online at https://liberationmath.wordpress.com/LGBTopinion/). Networks can also be visualized using many other programs such as Mathematica and Gephi.

Since the random graph used in the classroom will be quite small, certain things may need to be adjusted. Instructors should make sure there really are a wide range of opinion levels (a histogram is helpful here and can make a good visual for the class).

If there are nodes with a large number of connections, some may be trimmed to make the node easier for a student to handle.

The data from the network should then be used to populate node data sheets for the students (see Appendix A.2). This can be done by hand for a relatively small class, or by using a spreadsheet and merge program with a word processor to automatically produce the data sheets.

It is useful to have nametags for students that feature the node number on their data sheet and to set up the room to help students find each other by moving desks or tables out of the way. Student names and node numbers can also be posted in the class to help people to find each other during the simulation. It is also useful for the instructor to have access to a spreadsheet during class with some basic data about each node (level of acceptance and number of ties), which can be updated after the simulation is over.

3.2 During the Class Meeting. Having a student or course assistant help run the simulation will be a good idea, if possible. Once it is up and running, there are several things happening at the same time, and having an extra set of hands will be very helpful. New friend ties will need to be created during the simulation, so the instructor needs a method to create random ties during class, and it is useful to have an assistant get them out to the impacted students.

Each student will receive a data sheet that represents a random node (person) in a social network. This node does not represent the student themselves; for instance, the student may be bisexual, but the node may represent a person with a very low level of acceptance of same-sex relationships. It is useful to remind the students that these nodes do not represent their own identities and beliefs.

4 The Module

This module is designed to take one or two class periods, depending on how large the class is and whether additional follow-up and discussion happens in a second class period. This activity could also introduce a project which students would continue to work with on their own. The activity begins with an introduction to the issues of LGBT acceptance and to using graphs to model social networks. Then the class should review the procedures for the simulation, complete the simulation, and finally debrief the results.

4.1 Before the Simulation. Before the simulation, students need some discussion and visual aids that will help them understand the network and the process of opinion change and network adjustment. Students should understand that a graph can model a social network, and there are many easily accessible images of social networks that can be found with an image search of the words "social network" and "graph". The instructor could also draw an image of a volunteer student's social network. Key here is that networks can be modeled with graphs that are composed of nodes (people in the network) and edges (relationships).

The particular model the class will be using has two additional features. First of all, the nodes have an attribute,ranging from 0 to 1, corresponding to the level of acceptance of same-sex relationships. Lower values indicate less acceptance. A value of 1 is reserved for nodes that identify as gay, lesbian, or bisexual. This model also includes an attribute for each edge (relationship), indicating whether it is a friend relationship or a family relationship. The model then looks something like Figure 19.1. This model has a dynamic component in which nodes interact in a process that causes the changing of acceptance levels and adjusting of relationships.

4.2 The Simulation. Students should receive written instructions (see Appendix A.1) for the simulation, as the process is complex and involves multiple steps. Once the node information sheets and the student directions (both in the appendix) have been passed out, students will need to walk through a couple of examples as a group. A few minutes after starting the simulation, it may also be useful to pause it to address any questions that have come up.

Each interaction is initiated by one student. This student randomly selects a friend/family node to interact with by rolling a 12-sided die (or a 6-sided die if the student has fewer than 6 connections). Students need to know where on their data sheet to find their type of relationship, how to determine how far apart their acceptance level is from the selected friend/family node, and what the allowable differences are for friends and family. All of this is found in the student instruction sheet (see Appendix A.1).

If acceptance levels are close enough to each other, both nodes move closer to their average level using the process described in the student handout. Students should be reminded that they will be changing their acceptance level and they will bring their new acceptance level into any future interactions. The only exception to this is that LGB nodes never change their acceptance level, they remain at a constant level of 1.

If the acceptance levels are too far apart, the relationship is terminated, the edge is removed, and the node of the initiator is randomly assigned a new friend node, creating a new edge. It is important to note that **only** the initiator gets a new friend -- the other node does **not** get a new friend and they end the interaction with one less connection. This will have the impact over time of creating more isolation for nodes with very high or very low acceptance levels, but it may be good to let the students discover this for themselves. The instructor will need to use a random number generator to get initiators new friend nodes.

If the classroom has a document projector, it is useful to show the students how to put the information about interactions on their data sheets. Instructors should be clear with students that their acceptance level will be changing and that they bring their new acceptance level with them into any subsequent interactions. Instructors should also make sure that students note all interactions on their data sheets as well as their results.

Students should spend a fair amount of time having interactions, but there should also be time at the end of class to review results and to show students what happens in larger networks over time.

4.3 After the Simulation: Compiling Data. After the simulation, the instructor should collect all of the acceptance levels and number of ties for all of the nodes and display it back to the students, comparing it to the initial data. It might also be useful to color-code the acceptance levels so that it is easier to visualize the ways the network has changed. A capable student or assistant could work on compiling these while the instructor leads a class discussion. Alternately, the instructor can debrief the simulation with the students, but present the data at the next class. Histograms and information about standard deviation can also be useful, and could either be discussed in class or can provide an opportunity for follow-up activities by the students.

4.4 After the Simulation: Discussion. One good way to start the discussion is simply asking what students noticed as they did the simulation. As the discussion goes on, the instructor can bring in the voices of students who were managing nodes that were very accepting (LGB nodes or strong allies) or very unaccepting. These students will likely have experienced the breaking of a lot of ties. Be sure to discuss any changes in average acceptance level and numbers of ties.

It is also useful to have students discuss ways in which this does or does not reflect what happens in real life. If it doesn't come up naturally, instructors should bring up the parameters in the model, particularly the ones that control the strength of the ties. Changing these can have a big impact on the simulation, as can the overall percent of family ties. The discussion can also focus on how the model can be changed, which can lead to additional work that can be done outside of class (see Section 5).

4.5 After the Simulation: Computer Model. One limitation of the classroom simulation is the small number of interactions. To really get an idea of what happens over time, this needs to be done with a very large number of interactions, which means that it needs to be modeled with a computer. The instructor can show students images of the simulations below, or create other simulations and show those. The code for generating the simulations and additional color images of completed simulations can be found at `https://liberationmath.wordpress.com/LGBTopinion/`.

The images below are for a network of 50 nodes with 4000 interactions. A network that starts with a wide range of opinions and a few LGB people (see Figure 19.2) will end up progressing to a much higher average level of acceptance (see Figure 19.3). For instance, in the network of Figures 19.2–19.3, if we decide that a person would respond in a poll that they are "accepting" of people in same-sex relationships when their acceptance level is at 0.5 or above, then this network starts with 38% of the nodes accepting of same-sex relationships. However, at the end of the trial with 4000 interactions, 64% of the nodes are accepting. Note that it is the presence of LGB nodes that pulls the network to a higher level of acceptance since those nodes do not change their acceptance level and continue to interact with other nodes with lower acceptance level.

Other changes in the network can be seen in the network visualization of Figure 19.3. The LGB nodes, and some of their allies, are isolated from the rest of the network. They have fewer ties overall, and many have no ties with the dominant group, who are all clustered around an acceptance level around 0.5. There is a single unaccepting node remaining. In other trials of this simulation, unaccepting nodes can cluster together in a similar way. A network that starts with a higher average level of acceptance may not see this kind of clustering and isolation, and it is possible to obtain networks with several different clusters around different acceptance levels, whether or not LGB nodes are present in the network.

Another feature of this model is that family ties are very important. In the graph of the end of the trial, the ties connecting the allies with the LGB community are family ties (this can be seen more easily in color versions of these images which can be found at `https://liberationmath.wordpress.com/LGBTopinion/`), while the LGB community is well connected with friendship ties. This results because friendship ties are

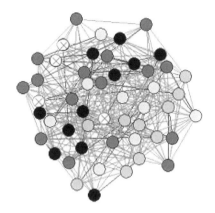

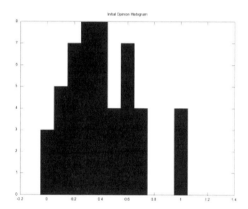

Figure 19.2. The starting network is on the top; the associated histogram of initial acceptance levels is on the bottom. Darker nodes are less accepting. Nodes with an "X" are LGB-identified. The acceptance levels in this network range from 0 to 0.8, except for the four nodes who are LGB (an acceptance level of 1)

relatively easy to break, but the LGB nodes are not able to easily make lasting relationships with the dominant community unless it is very accepting. Early on in the simulation many of the ties between LGB nodes and the dominant-group nodes are severed, and the LGB individuals are not as easily able to create new ties, since those ties get severed as well. It is important to note that this clustering remains even after the network becomes more accepting because the LGB nodes have found their own accepting community and have less interaction with the larger community. The stronger family ties can serve to pull the network to a more accepting state and can keep the LGB nodes connected with the rest of the community.

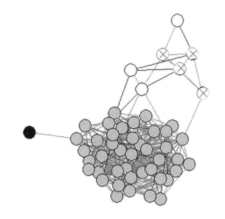

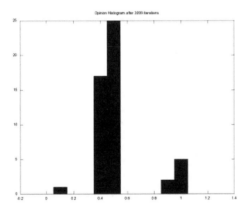

Figure 19.3. The network and the associated histogram of final acceptance levels after 4000 interactions. Darker nodes are less accepting. Note the clustering of acceptance levels into three distinct groups.

More trials and pictures can be found online at

https://liberationmath.wordpress.com/LGBTopinion/

as can the code to generate more networks and pictures.

5 Additional Thoughts

With a brief introduction to Octave or Matlab, students can make changes to the model and run additional simulations, even if they have fairly little programming experience. Students can experiment with the number of nodes, range of acceptance levels, the number of LGB nodes, the strength of the family and friend ties, and the percentage of the ties that are family. All of these have impacts on the final outcomes, and students can summarize results of the trials using histograms as well as network visualizations. Students can also consider what number of interactions are needed before the network becomes stable.

Students wishing to do a longer exploration could look at altering the implementation of the model. This modeling was done on a random graph, but there are better models of social networks, such as "small world" graphs. Students can also explore the impact of "coming out", in which a node might start with a disguised level of acceptance, but then reveal their true level of acceptance in an already stable network.

Appendix A: Assignments and Handouts

A.1 Modeling Change in Acceptance Using a Dynamic Graph.

Setup Instructions. The instructor will give you each a data sheet. This sheet represents a person, and contains a node number and a starting acceptance level. It also contains the ties or relationships that this node has.

Instructions for Each Round.

(1) You should have up to 12 nodes that you are linked with listed on your data sheet. At the beginning of the round, roll your 12-sided die. You will be interacting with the node that is listed next to this number on your data sheet (if you have no node listed in that spot, roll again). You are the initiator of the interaction. Record the node number on your data sheet.

(2) Find person with the node you have selected. Note that while you are trying to find this person, other people may be trying to find you. In other words, it will be a bit chaotic.

(3) Once you have found your partner, find the positive difference in your acceptance levels and compare it to the strength of your tie. Your tie is either a "friend tie" with strength 0.2 or a "family tie" with strength 0.9.

(4) One of two things will happen next:

    (a) If the difference between your two nodes is *greater* than the strength of your tie, then you will break the tie between the nodes.

        (i) *Both* of you should cross the tie off of your data sheets and record that the tie was broken.

        (ii) As the initiator in your interaction, you go now to the instructor and randomly receive another tie that you should write down on your data sheet.

        (iii) Note that your partner does *not* receive a new tie if your link is broken.

    (b) If the difference between your two nodes is less than the strength of your tie, then you will both change your acceptance levels.

        (i) Note: In our model, LGB nodes all have an acceptance level of "1" so LGB nodes will not change their acceptance; however their partners will change.

        (ii) Take the positive difference in your acceptance levels and *divide that by 4.*

(iii) The node with the *higher* acceptance level should *subtract* this amount from their acceptance level (but remember acceptance levels of 1 remain unchanged). The node with the *lower* acceptance level should *add* this amount to their acceptance level. (In this way your opinions get closer together.)

(iv) Record your new acceptance level on your data sheet.

(5) Once you are done:

(a) Check the board to see if you are a part of any newly-formed ties. If you are, record those in a blank spot on your sheet.

(b) Look around to see if anyone else is trying to initiate an interaction with you. If not, start over at #1 and initiate a new interaction yourself.

A.2 Node Data Sheet. Node #:
Starting Acceptance Level:
LGB or Straight:

Ties. (note that you may have fewer than 12 ties)

Tie #	Node	Friend or Family?
1		
2		
3		
4		
5		
6		
7		
8		
9		
10		
11		
12		

Interactions. *Note that LGB nodes always have an acceptance level of 1 and do not change

Node You Interacted with	Type of Tie (fr/fam)	Did you initiate? (yes/no)	Diff in acceptance level	Tie Broken? (yes/no)	New friend node	New acceptance level*

Bibliography

[1] Bronski, Michael. *A queer history of the United States.* Beacon Press, Boston, MA, 2011.

[2] Balazs Kozma and Alain Barrat, *Consensus formation on coevolving networks: groups' formation and structure*, J. Phys. A **41** (2008), no. 22, 224020, 8, DOI 10.1088/1751-8113/41/22/224020. MR2453832

[3] Gallup "In U.S., Record-High Say Gay, Lesbian Relations Morally OK" Retrieved October 14, 2014, from `http://www.gallup.com/poll/162689/record-high-say-gay-lesbian-relations-morally.aspx`

[4] Glicksman, E. "Transgender Today," *Monitor on Psychology*, 44:4 (April 2013) 36–36.

[5] M. E. J. Newman, *Networks*, Oxford University Press, Oxford, 2010. An introduction. MR2676073

[6] Newman, Mark E.J., Duncan J. Watts, & Steven H. Strogatz. "Random graph models of social networks." *Proceedings of the National Academy of Sciences*, 99.suppl 1 (2002): 2566-2572.

[7] Pew Research Center, "Growing Support for Gay Marriage: Changed Minds and Changing Demographics." Retrieved October 14, 2014, from `http://www.people-press.org/files/legacy-pdf/3-20-13%20Gay%20Marriage%20Release.pdf`

[8] Schelling, T. C. "Dynamic models of segregation." *Journal of mathematical sociology*, 1:2 (July 1971) 143–186.

20

Sustainability Analysis of a Rural Nicaraguan Coffee Cooperative

John Zobitz, Tracy Bibelnieks, and Mark Lester

Abstract. This module explores the varied aspects of sustainability in the context of a rural farming cooperative in Central America through analysis of resource allocation, revenue streams, and viability and long-term sustainability for ecotourism development. Specific applications of sustainability include bioeconomic models for development of ecotourism and optimization of financial resources for ecotourism marketing. The expert knowledge, perspectives, needs, and requirements of the diverse stakeholders (farmer, consumer, environment, government) are considered in the setup, analysis, and interpretation of mathematical models. The mathematical content is accessible to students in a first-year calculus or applied calculus sequence and includes rates of change, optimization, and modeling. **Keywords:** Bioeconomic sustainability, living wage, rates of change, modeling, project-based learning.

How do rural coffee farmers in Central America define and implement economic and environmental sustainability? What are the related social justice issues for the farmers and their families? Analyzing these questions in a calculus course provides both a real-world context to explore concepts of rates of change and infuses an international social justice perspective into the mathematics curriculum. Mathematics (and other natural science) programs are often rigidly sequenced with a prescribed progression from introductory to advanced coursework. This sequencing, while historical and necessary, is a barrier for mathematics students who desire exposure to international contexts in their education or coursework, especially as schedules become more constrained by specialized upper-level courses. Presenting international and applied contexts *early* in the first few semesters of the mathematics major provides motivation for advanced study of mathematics as well as planning for (additional) extended international opportunities. Furthermore, in adding an international dimension to the issues of sustainability and social justice, the context may enhance students' understanding of the complexities of these issues.

1 Mathematical Content

We present here a three-part module that can be used in most calculus classes within the course of three to four weeks, typically tied to a unit on the applications of the derivative. Students should have some knowledge of function families and algebraic solving skills developed in pre-calculus. If the module is given as a complete project, topics such as rates of change and their interpretation, differentiation, and optimization using differential calculus should already be familiar to students. As an alternative the module could be implemented in smaller pieces to motivate classroom discussion.

The specific mathematical topics and motivating questions addressed by the module are the following:

- **Modeling:** *What is a reasonable model for the number of daily visitors to the rural Nicaraguan coffee cooperative over the course of the year? What is the minimum price the cooperative can charge for a positive revenue?*

- **Parameter estimation:** *Given data provided by the cooperative and other sources, provide a reasonable estimate of parameters for your economic model.*

- **Related rates:** *Over the course of a year, how does the daily revenue of the cooperative change in relation to the number of visitors?*

- **Optimization:** *Use search engine analytics to determine the times of the year the cooperative should aggressively market itself to attract more visitors.*

- **Optimization:** *What is the number of visitors that maximizes the revenue?*

The module collectively strengthens the connection of the concept of rates of change to concepts from business and economics typically presented in calculus textbooks [5,10].

The learning outcomes we aim for include students being able to

- construct a mathematical model based on intuition and preliminary knowledge,

- incorporate eyewitness knowledge, and

- analyze data (student-collected or cooperative provided).

These learning outcomes align with some of the Cognitive and Content Recommendations listed in the 2015 CUPM curriculum guide [9], specifically:

- **Cognitive Recommendation 1:** Students should develop effective thinking and communication skills.

- **Cognitive Recommendation 2:** Students should learn to link applications and theory.

- **Cognitive Recommendation 4:** Students should develop mathematical independence and experience open-ended inquiry.

- **Content Recommendation 3:** Mathematical sciences major programs should include concepts and methods from data analysis, computing, and mathematical modeling.

We typically use this module to provide a real-world international context for our calculus classes. Alternatively, the module can be an entry point to advanced study of mathematical modeling. Outputs from the basic model are quantified in terms of a rate of change which naturally lends itself to ideas of accumulation and integration. Likewise, a basic differential equation model could be developed for customer acquisition and coupled to the revenue model. Additional reading on many of the concepts presented here can be found in [4].

2 Context / Background

Augsburg University is a private liberal arts university located in Minneapolis, Minnesota. The calculus sequence is a standard three-semester set of courses covering single-variable differential calculus, single-variable integral calculus, and multivariable calculus. In the first two semesters of the calculus sequence, students meet for class four days a week with one day held in a lab setting.

The module was developed through a partnership with the GARBO Coffee Cooperative, located in Peñas Blancas in northern Nicaragua. This area has beautiful white cliffs surrounded by local, family-owned coffee farms, but is relatively unknown to tourists. The cooperative was formed in 2000 to gain better access to the coffee market. In the 2011-12 cycle, farms of the cooperative produced 86,725 pounds of export-grade quality coffee for the world market. The GARBO Coffee Cooperative has certification from FLO-Cert, one of the fair trade certifiers in Latin America.

The predominant world economic policies tend to benefit large agribusiness involved in mono-cropping, to the detriment of the interests of small producers—a much larger group numerically who also tend to have more diversified farms. As a result more peasants are losing their land and rural poverty is on the rise. Organizing into cooperatives has been key for small producers to stay on their land, as they collaborate to buy inputs and sell their produce, achieving better economies of scale. Income diversification not only provides them with more sustainable livelihoods throughout the year, but also counteracts global warming, as they combine agricultural production with the preservation of the forests.

The GARBO Coffee Cooperative strongly supports the view that this diversification is possible without sacrificing the natural beauty of the location or adversely affecting the coffee crop. One option to meet these two goals (economic diversification and natural resource conservation) is farm-based ecotourism, and the cooperative has recently worked to develop their capacity in this area. Division of labor in cooperatives is mostly structured on traditional gender roles [8]. One natural way this diversification can be achieved is through engagement of women and youth. Such incorporation of additional groups in the economic livelihood of the cooperative may also lead to further democratization and participation in cooperative governance processes.

The background context for the students is quite minimal—it includes the description provided above as well as some two-minute videos of the cooperative and its members [1–3, 7]. The minimal context provided adds a sense of exploration and discovery into the process.

3 Instructor Preparation

3.1 Setting the stage. Setting the stage for the tasks is important for the success of the module. Hence, instructors considering adoption should learn first about the place and context and form their initial impressions from videos about the cooperative [1–3, 7]. Students will view these videos during the first part of the module as well. We also provide other specific information needed to complete tasks, put together in consultation with cooperative members.

Most standard calculus textbooks include a discussion of basic economic principles such as: *Revenue is the difference between profit and cost.* In such contexts, revenue is typically represented as a function of the number of items sold. In our context, the independent variable is v, or the number of daily visitors to the cooperative. If this profit model is studied in a unit on optimization, the goal is to determine the value of v, that is, the number of visitors, that will maximize the revenue function.

Standard linear revenue functions suggest that the revenue is maximized at the right endpoint of the interval for v; otherwise revenue grows without bound with increasing v for unconstrained optimization. In practical terms, a linear revenue function is unsustainable economically or environmentally in the long term. An alternative approach is using mathematical bioeconomics and bioeconomic optimization [4].

Two key concepts are needed to understand this approach. First, the revenue function will involve profit which we can describe by considering the rate of customer acquisition A (units person per day); when charged a price p (cordobas per person) these customers will bring in $p \cdot A$ cordobas per day to the cooperative. At the same time, the cooperative incurs two types of costs: *fixed* costs C, such as rent and utilities, that are independent of the number of visitors to the cooperative, and *variable* costs $k(v)$, which may decrease to a fixed value as the number of visitors increases due to efficiencies in scale. With these two assumptions, the revenue function then becomes

$$R(v) = p \cdot A - k(v) \cdot A - C.$$

The second concept is that the number of visitors v to the cooperative is a dynamic variable that depends on external interest in the cooperative, represented by a pool of potential customers. This customer pool could depend on the number of visitors to the cooperative. For example some visitors might not stay at the cooperative if it is too busy or overbooked. We define a yield function $Y = Y(v)$ with the properties that $Y(v)$ is zero when $v = 0$ and some other positive non-zero value when $v \neq 0$. A typical function form of Y is quadratic ($Y(v) = r \cdot v \cdot (M - v)$). Since the acquisition rate A also affects the visitor pool, the growth rate of the pool, labeled S for visitor stream, equals the difference between $Y(v)$ and A. Basic theory from economics suggests that in the absence of any other factors, the customer pool will be at equilibrium, which means that $S = 0$, or equivalently, that $A = S(v)$.

Bioeconomic optimization distinguishes between *maximum sustainable yield* v_M, the number of daily visitors where $Y(v)$ is optimized, and the *maximum economic yield* v_E, the value of v that optimizes the revenue function [4]. An illustration of these concepts is shown in Figure 20.1, with $Y(v)$ and $R(v)$ from the functions developed in the homework set we give to students. Since $v_E > v_M$, our model implies that if the number of daily visitors is at the level v_E, the cooperative will overextend their customer base, ultimately leading to a drop in revenue.

Yield function

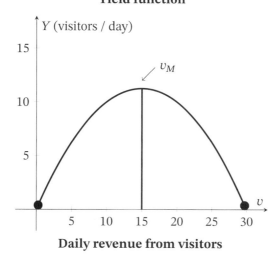

Daily revenue from visitors

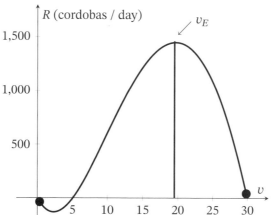

Figure 20.1. *Top plot:* Dependence of the yield function $Y(v)$ on the number of daily visitors. *Bottom plot:* Dependence of the cooperative revenue $R(v)$, considering fixed and variable costs. Since the maximum sustainable yield v_M is less than the maximum economic yield v_E, the cooperative is in danger of depleting their customer base over time if the number of daily visitors is at the level v_E.

Students will be using Google Trends (http://www.google.com/trends/) in Part 3 of the module (see Section 4.4), so instructors should probably familiarize themselves with the tool beforehand. Detailed instructions are provided in the appendix to guide students through obtaining data of interest from Google Trends. Furthermore, students will need to log in using a Google account, so it would be a good idea to encourage all students to get an account ahead of time.

3.2 Grading and evaluation of the student work. We designed the module so students would complete the work together but submit individual write-ups, so that we could assess individual student learning. To complement this, we recommend grading the module through a developmental approach. For example, students can submit work in pairs and then be asked to individually submit a typed written response to problems in addition to any pencil and paper work needed in their calculations. Shorter write-ups (no more than two pages) for each part of the module work best in terms of student willingness to complete, as well as efficiency in instructor grading. We recommend that each part of the module be graded using a simple rubric and the expectations of writing quality and point values increase over the duration of the project; for instance, instructors can assign 10 points for the first part, 20 points for the second, and 30 points for the third. We use the following instructions to convey expectations common to each write-up:

- **Audience:** You are writing an external consultant report for the cooperative.

- **Purpose**: One of your purposes will be to effectively communicate mathematical concepts and ideas to a non-technical audience.

- **Format of Project Report:**

    – 12 pt font, Times New Roman, paginated,

    – 1 inch margins, double spaced throughout,

    – begin body paragraphs with topic sentences,

    – include an introductory sentence describing the objective of this section,

    – conclude in a manner that gives the reader something poignant to think about (don't repeat the main points and avoid beginning the conclusion with "In conclusion").

4 The Module

The module can be spread out over the course of six weeks in a typical semester as a series of three different investigative tasks, along with a homework assignment on applied optimization, as shown in Table 20.1.

This timeline suggests that only towards the end of the project is the module topic directly aligned with calculus concepts. We prefer this gradual transition so we can stage students to be successful in applying mathematics outside of traditional textbook contexts.

In Appendix B, we provide a complete set of handouts which can be used for the whole module.

4.1 Part 1: Introduction, Fact-Finding, and the Revenue Model. A sample handout corresponding to this part of the module is available in Appendix B.1.

Table 20.1. Timeline of module tasks with approximate topics in a first-semester calculus course.

Weeks	Calculus Topics	Module topic
1-2	Instantaneous rate of change, tangent lines, and introduction to the derivative	Part #1: Introduction, Fact-Finding, and the Revenue Model
3-4	Differentiation rules	Part #2: Determining fixed and variable costs to accommodate visitors
5-6	Applications of differentiation	HW: Maximizing Revenue with Applied Optimization Part #3: Attracting Visitors to the Cooperative

Preparing students to investigate the module. The introduction to the module sets the context for the mathematical investigations of

(1) modeling and parameter estimation,

(2) application of related rates, and

(3) optimization.

The instructor should begin with an introduction to the coffee cooperative, including short videos produced for a student group that traveled to the region in previous years. Internet resources posted on the course website will encourage students to do further research on the fundamental geographic and socioeconomic background of the coffee cooperative. We recommend that the first assignment be for students to generate questions for the cooperative and determine what information can be found through the internet. Student groups of two to three work best for the research and preparation to engage with the module.

When we taught from these materials we asked the following sample questions in the first write-up:

- *After reviewing the websites and resources, what intrigued you about the initial fact-finding?*

- *What aspects of the cooperative do you want to know more about?*

- *If you were promoting travel to this region, what would you mention?*

- *What amenities would you want this region to have?*

Following this first part we found that students had a better general understanding of the scope of the project; we then introduced the basic revenue model to the students.

The basic revenue model. The basic model we use for revenue production R (where the units of R are dollars per day) is the following:

$$R = p \cdot v - C, \tag{20.1}$$

where we have the following variables:

> $R =$ daily revenue (\$ / day),
> $p =$ visitor charge (\$ / visitor) to stay at the cooperative,
> $v =$ the number of daily visitors (visitors / day),
> $C =$ the cost to the cooperative (\$ / day) to accommodate visitors.

Later it will be a good idea to change the monetary unit to cordobas, the local currency in Nicaragua, but at this point it is still reasonable to use the U.S. dollar (or whatever the local currency is) for familiarity. The basic model for revenue production is intentionally introduced without the traditional function notation (such as $R(p)$ or $R(v)$) to emphasize to students that each of the variables, p, v, C, represents a quantity that can change or, mathematically, is itself a function of time. While it is not necessary to emphasize at this point, by defining revenue as a quantity with units of dollars per day, i.e., as a rate, we could re-introduce the basic revenue equation as a derivative or differential equation.

The first exploration of the revenue model focuses on how changes in the quantities p, v, and C affect changes in daily revenue, the output of the model. Some guiding questions we ask students include:

(1) *Use several different values for p, v, and C to calculate revenue using the model. Interpret your results in each case.*

(2) *Examine equation (20.1) to determine an expression for the basic minimum price, p_{min}, the cooperative should charge to be profitable (i.e., $R > 0$). If the values of v and/or C change, how does the minimum profitable price change? Consider increasing or decreasing v and C independently and at the same time to answer this question.*

(3) *Now examine equation (20.1) to determine the minimum number of visitors, v_{min}, the cooperative must serve on a given day to be profitable. If both the price p and costs C change in value, what is the effect on the required minimum number of visitors?*

(4) *Assuming the price charged to each of the visitors is fixed, discuss with your group some strategies that the cooperative can pursue to potentially boost revenue. Make sure to come up with at least three strategies. Evaluate the three on their merits, and determine appropriate data and metrics in your evaluation. With your group, decide on the optimal strategy that you would recommend moving ahead with, providing justification in your report to the cooperative.*

4.2 Part 2: Determining fixed and variable costs to accommodate visitors.
In this part, we revisit the basic model for revenue production to engage students in the specific context of the coffee cooperative. The relevant part of our sample handout is in Appendix B.2.

A revised model. Students are asked to use the internet to determine the currency used in Nicaragua (cordobas) and research the conversion between their currency and Nicaraguan cordobas. Knowing this conversion is essential for students to make sense of their numerical results. Students are then given a table of common goods and costs (Table 20.3 in Appendix A) that the cooperative makes use of both in their daily living and in providing for tourists.

At this point we introduce a revised revenue model that modifies the basic revenue model in equation (20.1) by separating out the cost per visitor from the fixed costs, C:

$$R = p \cdot v - k \cdot v - C. \qquad (20.2)$$

Here $k =$ visitor cost (cordobas / visitor) to accommodate guests. Equation (20.2) refines the revenue analysis to allow for consideration of differences between visitors or groups of visitors. Moreover, this model allows for analysis where costs per visitor are affected by external factors such as season, availability of goods, or the cooperative members' ability to travel to purchase goods.

When analyzing equation (20.2), the students' goal is to estimate the daily cost (in cordobas) to accommodate visitors using research and/or the information provided by the instructor. We guide students into this exercise by asking them to plan out typical meal requirements for a given guest and create an itemized list of items. We encourage them to include items they think are important that are not listed in the table but are required to provide a justified reasonable estimate of price(s). Students are then asked to identify:

(1) costs that would be dependent on each visitor and

(2) costs that would be fixed or visitor-independent.

Finally, students answer a series of guiding questions to relate their work back to equation (20.2), focusing on the parameters k and C.

An opportunity for in-class discussion. At this point of the module there is room for some reflective in-class discussion. The table given to students (Table 20.3 in Appendix A) is a true reflection of the simplicity of life in the cooperative. For example, carbonated beverages, red meat, many common vegetables, breads, and cereals are not listed in the table. This may be a good opportunity to encourage students to consider their own diets and use of consumable goods. Some guiding questions for discussion could be:

- *What food and goods do you consume or use on a daily basis?*

- *How is life in your circumstances different from that of the cooperative?*

Students may further research the area of Nicaragua where the cooperative is located. Some guiding questions could be:

- *What food and consumable goods are readily available and which are cost prohibitive?*

- *How might this negatively affect the opportunity for ecotourism or how might it be leveraged to increase interest by visitors?*

Connecting to calculus concepts. We time the conclusion of the second part to coincide with our discussion of the chain rule and of related rates. We aggregate student results to come up with estimates of p, k, and C, and present possible periodic functions that would represent seasonal patterns in the number of visitors over the course of the year. One periodic function that we developed in this manner was

$$v(t) = -10 \cos\left(\frac{2\pi t}{365}\right) + 10.$$

Annual daily revenue

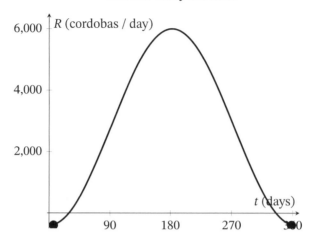

Figure 20.2. Time dependence of the cooperative revenue throughout a year, given a conceptual model for the number of visitors. The revenue is computed via equation (20.2), with $p = 670$, $k = 350$, $C = 400$, and $v(t) = -10\cos\left(\frac{2\pi t}{365}\right) + 10$.

The resulting revenue pattern, obtained via equation (20.2), is shown in Figure 20.2.

The discussion on related rates can then center around annual graphs of the revenue as a function of time. This part could be concluded with a reflective assignment or a group discussion to evaluate the feasibility of the revenue model for the cooperative in terms of the other ongoing work, such as the coffee harvest.

4.3 Homework: Maximizing Revenue with Applied Optimization. This part of the module is a transition from the refined revenue model to thinking about how the cooperative will attract customers. The first approach builds on the pool of visitors and the yield function $Y = Y(v)$. It is convenient to use the notation $Y(v)$ (with units of people / day) because the function $Y(v)$ can be interpreted as a yield or supply function in an economic context [4]. A key mathematical assumption is that $Y(v)$ is zero when $v = 0$ and some other positive non-zero value when $v \neq 0$. A typical function form that fits this requirement would be the quadratic. For instance, we can use $Y = r \cdot v \cdot (M - v)$, where we have the following parameters:

$r =$ rate that knowledge about the cooperative spreads per day,
$M =$ maximum number of people who know about the cooperative.

If we then define A to be the rate that potential customers visit the cooperative, then we have:

$$S = r \cdot v \cdot (M - v) - A, \tag{20.3}$$

where S stands for the visitor stream, or the growth rate of the customer pool.

The in-class discussion at this point could center on several different conceptual graphs of $Y(v)$ with A, connecting the sign of S to growth/decline of the potential customer pool. This discussion is similar to the discussion of the basic revenue model in equation (20.1). Students can also determine the equilibrium values of the model (that is, when the customer pool is not changing or when $S = 0$, which should easily demonstrate that equilibrium is when $A = Y(v)$). The objective of this analysis is for students to use the cooperative as a context for the graphical relationship between a function and its derivative. If students can use the stream S, a rate of change, to reach some conclusions about the changing quantity itself (the pool), then this can be a productive discussion.

More generally, we can write equation (20.3) as:

$$S = Y(v) - A, \tag{20.4}$$

and couple it with a more general form of the revenue function:

$$R(v) = p \cdot A - k(v) \cdot A - C. \tag{20.5}$$

Recall that here

$p =$ price charged per visitor (cordobas / visitor),
$k(v) =$ cost to accommodate visitors (cordobas / visitor),
$C =$ the fixed cooperative costs (cordobas / day) to accommodate visitors.

In particular we generalize k, a constant in the linear model of equation (20.2), to be a function of v.

Basic theory from economics suggests that in the absence of any other factors, the customer pool will be at equilibrium, that is, it will remain constant. This will only happen when the customer stream, S, equals zero, or equivalently, when $Y(v) = A$. Under this assumption, students can now investigate the relationship of this economic model to applied optimization in a set of guided homework problems examining the value of v that maximizes $Y(v)$ (the *maximum sustainable yield*) and the value of v that optimizes $R(v)$ (the *maximum economic yield*). A handout for this homework assignment is available in Appendix B.3.

4.4 Part 3: Attracting Visitors to the Cooperative. This part utilizes Google Trends, a data analytics site that allows students to explore publicly accessible aggregated search data from across the world, to investigate models for the rate of change of visitors to the cooperative. In a business setting, data analytics are extremely relevant; for instance, use of data analytics is a driving force behind development of effective social media marketing campaigns.

To use Google Trends (`http://www.google.com/trends/`) students log in using a Google account. Students then type in a search term relevant to the cooperative that might be used in Google by people interested in an ecotourism experience like that offered by the cooperative. For each search term typed into Google Trends, a plot is generated by Google that shows the relative interest (scaled from 0 to 100) in the search term over the time period specified by the search. We leave it open to students to decide a particular search term but justify why it could be relevant for the cooperative. For any search term, the data from the plot of relative interest over time can be downloaded as a comma separated (csv) file that can be opened in a spreadsheet application such as Excel or imported into a statistical software program (see Figure 20.3 for a typical plot obtained in this way). Students then apply function fitting capabilities of the spreadsheet or other software to explore possible functions that could be used to model the data.

After this activity we aggregate students' results and their recommendation for specific search terms to the cooperative as well as times of the year where it would be

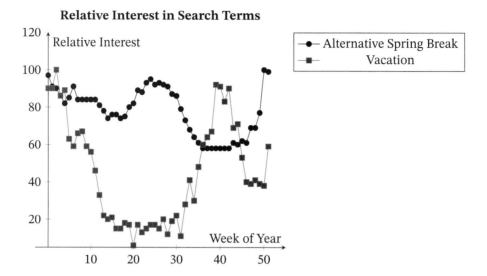

Figure 20.3. Relative interest in internet searches with the terms "Alternative Spring Break" (round blue dots) or "Vacation" (solid red squares) as determined from Google Trends. Higher relative interest indicates more searches relative to other time periods. Note that there is a cyclical pattern to the search terms shown (this cyclical behavior is apparent when viewing search data over multiple years). The cooperative could consider purchasing internet advertising timed during peak periods of interest to generate more visitors to the cooperative.

more advantageous to purchase a sponsored advertisement connected to a particular search term (Table 20.2). If the cooperative created a sponsored internet advertisement tied to when people search for "vacation" in Google, it might generate more visitors to the cooperative.

Table 20.2. Specific search terms from Google trends and their corresponding time periods where it would be more advantageous to purchase a sponsored advertisement connected to a particular search term.

Search Term	Times of year when people search
Nicaragua tourism	January - March
Coffee cooperative	August - January
Vacation	March - August, November - December
Break	November - December
Alternative spring break	January - March, October - December
Coffee	January - February

5 Additional Thoughts

5.1 Application notes. When we taught this material, students determined the minimum charge per guest to ensure profitability in Part 1 of the module. After aggregating the class results together, we were able to recommend a minimum charge and a range of potential profitability for the GARBO Coffee Cooperative. Second, as a class we recommended a variable stipend amount to host families, to provide a supplemental source of income. Finally, analysis of internet marketing data from Google Trends provided information when specific search terms (such as *Nicaragua tourism*, *vacation*, *coffee cooperative*) were more likely to occur. The cooperative could then consider purchasing internet ads linked to those search terms as a way to generate interest in the cooperative. As instructors for the module, we wrote these recommendations in an executive summary to be shared with the cooperative. Participating in the writing of such an executive summary of recommendations could demonstrate to the class the value and real benefit of the results from engaging with the project. Thus as an alternative, students could be asked to write the summary.

We implemented this module as part of a first-semester calculus course in the fall of 2014. All of the student recommendations were presented to the cooperative leadership for further deliberation at the completion of the semester. Following the presentation of results, the cooperative was motivated to do more research. With support from the Winds of Peace Foundation (an international non-profit foundation, `peacewinds.org`) women in the cooperative are doing a more detailed investigation into costs and profitability. The cooperative is not in a position for internet marketing at this time but might pursue such an option in the future. Based on this work, each year the cooperative defines a project for students to analyze; in 2015 students examined the carbon footprint of a cooperative farm utilizing remote sensing data (`https://lpdaac.usgs.gov/`).

There are several benefits to engaging students in such a project. Both the students and the cooperative are equal partners in the learning process, creating more opportunities for reciprocal understanding [**6**]. We believe that students are empowered to pursue additional study of mathematics and also interested in tying their study with international education. The cooperative has a richer context to explore and investigate different options pertinent to their future economic sustainability and success. Finally, as instructors we are enriched by the connections and cross-cultural engagement and learning facilitated through this module. We invite readers to contact us if they are interested in participating in this ongoing project.

5.2 Extending the module. There are several extensions that could follow the homework set in Section 4.3. Students could complete a reflective assignment or group discussion on the differences between the maximum sustainable yield and the maximum economic yield. Other optional extensions are simulation of the coupled system of equations with student developed models for customer acquisition A and effect on the annual revenue for the cooperative.

5.3 Possible variants. We recommend implementing the module in smaller pieces (three parts over the course of the term) rather than as one large project together. This helps distribute the grading and provides an opportunity for students to develop their skills in writing and preparation of the reports. The module, however, could definitely serve as an independent study project over the course of the term.

An alternative approach is to consider the guiding principles for each part throughout as a motivating example for particular calculus concepts. For example the basic model has outputs as a rate of change (currency per day or visitors per day). Introduction of the revenue model as the difference between profit and costs along with some conceptual graphs provides a contextual example connecting the sign of the derivative to function properties (i.e., increasing, decreasing, extrema, concavity). Implementation of the module in this way provides a unified theme, a story tying together the different mathematical ideas developed over a traditional semester of calculus.

Acknowledgments. Kernels of this project began when John Zobitz participated in an undergraduate study abroad experience in Nicaragua, followed by an interest to integrate some of those experiences into mathematics courses. Augsburg University has had a campus in Nicaragua since 1984. Our initial experience with the cooperative was during a short-term study abroad program in 2012 [11]. Funding for this work was supported through an Augsburg University Integrated Course Design Grant and for Engaging Mathematics, with support from the National Science Foundation. *Engaging Mathematics* is an initiative of the National Center for Science and Civic Engagement. Updates and resources developed throughout the initiative are available online at `www.engagingmathematics.net`. John Zobitz thanks N. Schoenberg for support and feedback on earlier drafts of this manuscript. The authors thank two anonymous reviewers, whose feedback greatly improved this manuscript.

Appendix A: Tables and Datasets

Here is a table of common costs of things needed to accommodate guests from a visit to the cooperative in March 2012, as provided by cooperative leaders:

Table 20.3. Common costs of things needed to accommodate guests from a visit to the cooperative in March 2012.

Item	Cost	Notes
Food		
Rice	10 cordobas / pound	1 pound feeds six people
Beans	8 cordobas / pound	1 pound feeds six people
Oil	18 cordobas / 0.25 Liters	Used for cooking
Tortillas	2 cordobas / 1 pound corn	1 batch makes 10 tortillas
Chicken	120 cordobas / hen	
Eggs	3 cordobas / egg	
Mango	10 cordobas	
Watermelon	100 cordobas	
Cantaloupe	40 cordobas	
Bread	29 cordobas	Price for 1 loaf
Pasta	8-10 cordobas	1 package feeds six people
Sugar	9 cordobas	
Cheese	35 cordobas	
Coffee	Cost of labor	
Purified water	20 cordobas / 500 mL	Needed for food cooking and consumption
Consumables		
Toilet paper	12 cordobas / roll	
Napkins	35 cordobas / package	A package has 60
Candles	2 cordobas / candle	1 candle lasts approximately 1.5 hours
Body soap	14 cordobas / bar	
Laundry soap	16 cordobas / load	
Matches	1 cordoba / box	The cooperative does not have a reliable source of power.
Batteries	25 cordobas / battery	

Additional market prices for items can be found at the following website: http://www.numbeo.com/cost-of-living/country_result.jsp?country= Nicaragua. To use this website for this module, readers should note that the nearest city to the cooperative is Matagalpa.

Appendix B: Assignments and Handouts

Here we provide a handout packet that can be used for the complete module.

Sustainability Analysis of a Rural Nicaraguan Coffee Cooperative

How do rural coffee farmers in Central America define and implement economic and environmental sustainability? If this sustainability is not achieved, what are the social justice issues for the farmers and their families? What role does government play in either supporting both economic and environmental sustainability? Through answering these questions you will analyze multifaceted aspects of sustainability of a rural farming cooperative in Central America through analysis of resource allocation, revenue streams, and viability and long-term sustainability for ecotourism development.

A coffee cooperative in the Peñas Blancas region of Nicaragua has hired you as an external consultant to evaluate opportunities for additional revenue generation to offset any losses due to shortcomings in the coffee crop. In order to evaluate this, you will need to determine the relevant factors and sources for alternative revenue streams.

B.1 Part 1: Introduction, Fact-Finding and the Revenue Model.

Learning more about the cooperative. Below are several sources (in English) describing the region we are focusing on.

Internet resources on the Peñas Blancas region:

- http://www.lonelyplanet.com/nicaragua/la-dalia-penas-blancas

- http://www.nicaragua.com/blog/visiting-the-spectacular-penas-blancas-massif

- https://vianica.com/go/specials/26-penas-blancas-massif-nicaragua.html

YouTube videos of the cooperative:

- https://www.youtube.com/watch?v=rOr8bhIqgzs

- https://www.youtube.com/watch?v=j4pkwNpxY2I

- https://www.youtube.com/watch?v=sLRbrL_0TRg

- https://www.youtube.com/watch?v=pXUklmPVIrE

Read through these websites with your group and discuss the following questions:

(1) What intrigued you about the initial fact-finding?

(2) What do you want to know more about?

(3) If you were promoting travel to this region, what would you mention?

(4) What amenities would you want this region to have?

The basic revenue model. A basic model for revenue production R (units dollars per day) is the following:

$$R = p \cdot v - C, \tag{20.1}$$

where we have the following variables:

$$
\begin{aligned}
R &= \text{ daily revenue (\$ / day),} \\
p &= \text{ visitor charge (\$ / visitor) to stay at the cooperative,} \\
v &= \text{ the number of daily visitors (visitors / day),} \\
C &= \text{ the cost to the cooperative (\$ / day) to accommodate visitors.}
\end{aligned}
$$

Considering equation (20.1), answer the following questions:

(1) Let's say the daily rate for the cooperative is $20 per person, and it costs $75 per day for upkeep. What is the revenue if the cooperative has five visitors? What about one visitor? How would you interpret these results?

(2) Instead of focusing on specific values, examine equation (20.1) to determine an expression for the basic minimum price for the cooperative to be profitable. If the number of visitors v and the cost C change in value, what does increasing or decreasing both of them do to the minimum price?

(3) Related to the last point, examine equation (20.1) to determine the minimum number of visitors for the cooperative to be profitable. If both the price p and the cost C change in value, what does increasing or decreasing both of them do to the required minimum number of visitors?

(4) Assuming the price charged to each of the visitors is fixed, discuss with your group some strategies that the cooperative can pursue to potentially boost revenue. Make sure to come up with at least three strategies. Evaluate the three strategies on their merit, and determine appropriate data and metrics in your evaluation. With your group, decide on the optimal strategy that you would recommend moving ahead with. Provide justification in your report to the cooperative.

Your Report for Part 1: While writing your report for Part 1, keep the following in mind:

- **Audience:** You are writing an external consultant report for the cooperative.

- **Purpose**: One of your purposes will be to effectively communicate mathematical concepts and ideas to a non-technical audience.

- **Format of Project Report:**

    - 12 pt font, Times New Roman, paginated,
    - 1 inch margins, double spaced throughout,
    - begin body paragraphs with topic sentences,
    - include an introductory sentence describing the objective of this section,
    - conclude in a manner that gives the reader something poignant to think about (don't repeat the main points and avoid beginning the conclusion with "In conclusion").

- **Content:**

    - Your introductory fact-finding results, impressions, and questions for the cooperative.
    - Description of the basic model and explanation of different strategies the cooperative can do to increase their revenue.
    - Summary of the strategies you discussed with your group, with justification that the optimal strategy with appropriate data and metrics in your evaluation. Be sure to cite any consulted sources.

B.2 Part 2: Determining fixed and variable costs to accommodate visitors.
In this part of the project we will revisit the basic model for revenue production, using key aspects of how the revenue is defined. As a reminder, a basic model for revenue production R (units dollars per day) is the following:

$$R = p \cdot v - C, \tag{20.1}$$

where we have the following variables:

$R =$ revenue / day ($ / day),
$p =$ visitor charge ($ / visitor) to stay at the cooperative,
$v =$ the number of daily visitors (visitors / day),
$C =$ the cost of the cooperative ($ / day) to accommodate visitors.

(1) The currency in Nicaragua is the cordoba. Identify the current conversion between United States dollars and the Nicaraguan cordoba. We will now write all the currency units in Nicaraguan cordobas, so knowing this conversion will be helpful to think about your results in a familiar currency.

(2) With your group, estimate for the daily cost to accommodate visitors using the information provided. **Use the Nicaraguan cordoba as the monetary unit.** It might be helpful to plan out a typical meal requirement for a given guest and to create an itemized list. Feel free to include items you may think are important but are not listed, and provide a reasonable estimate on the price (with justification!). Justify your reasoning in determining the items that would be important.

(3) Go back to the itemized list you made. Identify costs that would be (1) dependent on each visitor and (2) costs that would be fixed or visitor-independent.

Based on this information, we can refine our basic model to the following:

$$R = p \cdot v - k \cdot v - C, \tag{20.2}$$

where we have introduced the following variable:

k = visitor cost (cordobas / visitor) to accommodate guests.

(4) Assuming the values of p, k, and C are known, what is an algebraic expression for the minimum number of visitors needed for the cooperative to be profitable? How does changing the parameters p, k, and C affect the number of visitors?

(5) Examine your algebraic expression from the last problem. What is a necessary condition for the value of p?

(6) Based on your itemized lists of fixed and variable costs, estimate C. As you can imagine, the value of the parameter C depends on which items you focused on in Table 20.3 as visitor-dependent. Provide justification for your reasoning.

(7) Based on your itemized lists of fixed and variable costs, estimate k (with justification). To determine k, it might be useful to take an accounting of the per item cost of goods, estimate the number of items that each visitor would use (for every day or for a visit), and then multiply the two values together. From that work determine if a summative or an average cost per visitor is worthwhile.

(8) Now that you have determined C and k, what is a necessary condition for p? Would that price p be reasonable for visitors to the cooperative?

(9) A recommendation from an external group is to give a daily wage to each family to offset the cost of receiving visitors. The wage should incorporate the fixed cost used to accommodate guests and also any (human) time spent in the receiving of guests. If the cooperative will implement this plan, what type of information would be useful in setting a wage? What would be the wage you would recommend? Based on your investigations, determine how this wage will affect p, the price they charge to visitors. Would that price p be reasonable for visitors to the cooperative?

Your Report for Part 2. While writing your report for Part 2, keep the following in mind:

- **Audience:** You are writing an external consultant report for the cooperative.

- **Purpose**: One of your purposes will be to effectively communicate mathematical concepts and ideas to a non-technical audience.

- **Format of Project Report:**

    - 12 pt font, Times New Roman, paginated,

    - 1 inch margins, double spaced throughout,

    - begin body paragraphs with topic sentences,

    - include an introductory sentence describing the objective of this section,

    - conclude in a manner that gives the reader something poignant to think about (don't repeat the main points and avoid beginning the conclusion with "In conclusion").

- **Content:**

    - Your introductory fact-finding results, impressions, and questions for the cooperative.

    - Description of the basic model and explanation of different strategies the cooperative can do to increase their revenue.

    - Summary of the strategies you discussed with your group, with justification of the optimal strategy with appropriate data and metrics in your evaluation. Be sure to cite any consulted sources.

    - Your determination of which costs were important, fixed costs, and visitor-dependent costs.

    - Description and justification of the parameters p, k, and C, and how you set the minimum price p.

    - Evaluation of the minimum wage you would give to the families.

    - Evaluation of the feasibility of the price for visitors. Be sure to cite any consulted sources.

B.3 A homework assignment: Maximizing Revenue with Applied Optimization. **Instructions:** Answer the following questions on a separate piece of paper, numbering each problem and using careful mathematical notation.

A basic revenue model. We have seen that a basic model for revenue production R (where the units of R are dollars per day) is the following:

$$R = p \cdot v - k \cdot v - C, \tag{20.2}$$

where we are using the following variables:

$R =$ revenue / day (cordobas / day),
$p =$ price or visitor charge (cordobas / visitor) to stay at the cooperative,
$v =$ the number of daily visitors (visitors / day),
$k =$ visitor cost (cordobas / visitor) to accommodate guests,
$C =$ the cost of the cooperative (cordobas / day) to accommodate visitors.

Here are the specific questions you should address:

(1) Consider the revenue equation $R(v) = 670v - 300v - 300$. What is the value of v that maximizes R over the interval $0 \le v \le 30$?

(2) Examine equation (20.2) to determine an expression for the basic minimum price, p_{min}, the cooperative should charge to be profitable (i.e., $R > 0$). By examining your expression, if the values of v and/or C change, how does the minimum profitable price change? Consider increasing or decreasing v and C independently and at the same time to answer this question.

(3) Now examine equation (20.2) to determine the minimum number of visitors, v_{min}, the cooperative must serve on a given day to be profitable (i.e., $R > 0$), this time treating p and C as constants. By examining your expression, if both the price p and costs C change in value, what is the effect on the required minimum number of visitors? Consider increasing or decreasing p and C independently and at the same time to answer this question.

A revised revenue model. More generally, we assume that there is a pool of potential visitors to the cooperative. This customer pool may fluctuate as interest in the cooperative increases or wanes (for example, some visitors might not stay at the cooperative if it is too busy or overbooked).

Assume that the cooperative has a customer acquisition rate A (units visitors/day). A basic model that represents the **growth rate of the customer pool**, that is, the customer stream is:

$$S = Y(v) - A, \tag{20.4}$$

where $Y = Y(v)$, with units of people / day, represents a customer growth function which can be interpreted as a yield or supply function in an economic context. A key mathematical assumption is that $Y(v)$ is zero when $v = 0$ and some other positive non-zero value when $v \ne 0$.

In the absence of any other factors, we have an equilibrium when the pool size is not changing, or equivalently, when the customer stream is zero. This corresponds to $Y(v) = A$. This equilibrium assumption modifies the basic revenue model (equation (20.2)) to the following:

$$R(v) = p \cdot A - k(v) \cdot A - C, \tag{20.5}$$

or equivalently,

$$R(v) = (p - k(v)) \cdot Y(v) - C. \tag{20.6}$$

Under this assumption we can now investigate the relationship of this economic model to applied optimization. We will need two new definitions. The value of v that maximizes $Y(v)$ is called the *maximum sustainable yield*. The value of v that optimizes $R(v)$ is called the *maximum economic yield*.

With that context, answer the following questions:

(1) What relationship between p and $k(v)$ will be profitable? Why?

(2) Consider the following customer growth function $Y(v) = 0.05v \cdot (30 - v)$. Determine the value of v that maximizes $Y(v)$.

(3) More generally, consider the function $Y(v) = r \cdot v \cdot (M - v)$, where r and M are constants. For this equation, determine the value of v that maximizes $Y(v)$. Your final result should be an expression involving r and M. Check your work using the values of $r = 0.05$ and $K = 30$. Does it match what you found in the previous problem? Call this value of v the *maximum sustainable yield*.

(4) An equation for the per visitor cost is $k(v) = 100 \cdot 0.85^v + 200$. Sketch this function over the interval $0 \le v \le 30$. Give a contextual explanation for the behavior of this function, as compared to a fixed constant cost function of $k(v) = 300$.

(5) Graph the general revenue equation $R(v) = (670 - 100 \cdot 0.85^v - 200)(0.05v \cdot (30 - v)) - 300$ using graphics software. Determine on the graph the value of v that maximizes the revenue function $R(v)$. Call this value of v the *maximum economic yield*. How does this compare to the value of the *maximum sustainable yield*, the value of v that maximizes $Y(v)$?

(6) Use calculus to determine an expression for $R'(v)$. Try to solve $R'(v) = 0$ to determine the *maximum economic yield*.

(7) Theory from mathematical bioeconomics suggests that if the maximum economic yield is greater than the maximum sustainable yield, then the resource (in this case visitors to the cooperative) is over-extended, leading to an ultimate decline in visitors. What are the implications for this result in terms of the long-term sustainability of the cooperative?

B.4 Part 3: Determining the number of visitors to the cooperative. The cooperative would like to know how to market themselves to attract the desired number of daily visitors (about 15-20 people per day) in order to maximize their revenue. In order to do this, we are going to examine reported data from search engines such as Google. The website that we will be using is Google Trends, a data analytics site that allows you to explore aggregated search data from across the world.

Please follow the following steps to start your data analysis.

(1) Google Trends is a website that allows you to analyze data used in Google search engines. In order to use this you will need to log in with a Google account.

(2) Navigate to Google Trends `http://www.google.com/trends/`

(3) Type a search term in the topics menu bar. A plot should be generated highlighting the relative interest (scaled between 0 to 100) in this search term over time.

The data are scaled in this way so searches in countries with more active internet users would not dominate the results. If you are curious for more information about how the data are scaled, please visit the following link: `https://support.google.com/trends/answer/4365533?hl=en&rd=1`

(4) If you are comfortable now with the interface for Google Trends, select a search term that will be relevant for the cooperative in terms of their marketing (e.g., knowing how Google users trend for "Game of Thrones" is not as helpful as "winter escape"). You might have to try a few different terms in order to get some data. Record terms that were useful and not as useful.

(5) For the purposes of this project, we are going to limit the results to one year. To do that, click on "select time range" and limit it to **one complete calendar year**.

(6) Now let's download the data. At the top right of the webpage you should see a gear icon, allowing you to download the data as file "Download as CSV". When you do this, you will download a **c**omma **s**eparated **v**alues file called "report.csv". Save this csv file and open it. When you open up this file in a spreadsheet program, the first few lines provide information about the web search. Starting at about line 6 is the "interest" data graphed on the webpage. Notice that in column "A" there is a time stamp corresponding to the weeks. We will rewrite this as weeks of the year.

  (a) First, create a column labeled Weeks as follows:

  (b) In cell A6 type the following formula in quotations: "=1".

  (c) In cell A7 type the following: "=A6+1".

  (d) Now, click and drag the lower right corner of cell A7 down the length of the column. This will autofill the formula down. The last cell, A57, should read 52.

(7) Now create a scatter plot for the Interest data over the course of the year. You will want to select the column of the weeks of the year and the interest. Make a scatter plot of the data.

(8) Title your chart and axes appropriately.

(9) Let's try to fit different function families to these data. We will let I represent the interest in your search term and t represent the weeks since January 1, so you are determining a functional form $I(t)$. For each of the function families listed below, fit a curve through the data, and for each one, **record how good your model is in representing the data**:

  (a) linear formula:

  (b) quadratic:

  (c) cubic:

  (d) exponential:

  (e) logarithmic:

(10) If one was to try to fit a curve that represented the data the best, which function family would you choose and why?

(11) With the modeled function that you chose, now use your calculus knowledge to determine the following:

    (a) Intervals where $I(t)$ is increasing.

    (b) Intervals where $I(t)$ is decreasing.

    (c) Times of the year when $I'(t)$ is zero ($I(t)$ has a critical point).

    (d) Times of the year when $I(t)$ is a maximum and a minimum.

(12) What does $I'(t)$ represent mathematically? How would you describe it contextually?

(13) Go back to the intervals you identified above. What would they mean in the context of marketing the cooperative?

(14) Based on your analysis, during what times of the year should the cooperative consider online and social media marketing in order to attract more customers?

Your Report for Part 3. While writing your report for Part 3, keep the following in mind:

- **Audience:** You are writing an external consultant report for the cooperative.

- **Purpose**: One of your purposes will be to effectively communicate mathematical concepts and ideas to a non-technical audience.

- **Format of Project Report:**

    – 12 pt font, Times New Roman, paginated,

    – 1 inch margins, double spaced throughout,

    – begin body paragraphs with topic sentences,

    – include an introductory sentence describing the objective of this section,

    – conclude in a manner that gives the reader something poignant to think about (don't repeat the main points and avoid beginning the conclusion with "In conclusion").

- **Content:**

    – Your choice of search term in Google Trends with justification.

    – A graph of the interest data over the course of the year, along with your fitted function (from Excel).

    – A summary of the model-data fitting with the different function families.

    – An in-depth explanation and justification of which function family you chose to represent the data.

    – A graph of $I'(t)$ (can be done using Mathematica or other appropriate software).

    – A non-technical explanation of the meaning of $I'(t)$ and identification of the intervals and extreme values listed above, along with their interpretation in a non-technical context.

    – An evaluation of when the best time during the year is for the cooperative to consider online social media marketing to attract more business.

Bibliography

[1] Augsburg Center for Global Education & Experience. *Calculus in Nicaragua*. `https://www.youtube.com/watch?v=j4pkwNpxY2I`. Online; accessed 2016-04-09.

[2] Augsburg University. *The GARBO Coffee Cooperative*. `https://www.youtube.com/watch?v=sLRbrL_OTRg`. Online; accessed 2016-04-09.

[3] Augsburg University. Mark Lester Interview. `https://www.youtube.com/watch?v=pXUklmPVIrE`. Online; accessed 2016-04-09.

[4] Colin W. Clark, *Mathematical bioeconomics*, 3rd ed., Pure and Applied Mathematics (Hoboken), John Wiley & Sons, Inc., Hoboken, NJ, 2010. The mathematics of conservation. MR2778605

[5] Deborah Hughes-Hallett, Karen R. Rhea, Andrew Pasquale, Andrew M. Gleason, William G. McCallum, David O. Lomen, David Lovelock, Jeff Tecosky-Feldman, Thomas W. Tucker, Daniel E. Flath, and Joseph Thrash. *Calculus: Single Variable*. Wiley, 5th edition, February 2010.

[6] Peter Levine and Karol Edward Soltan, editors. *Civic Studies*. The Civic Series. January 2014.

[7] Mark Lester. *GARBO Families with Augsburg Mathematics*. `https://www.youtube.com/watch?v=rOr8bhIqgzs`. Online; accessed 2016-04-09.

[8] René Mendoza. *The Associativity Route as a Driver of Development: Study on Cooperatives in the Central Northern area of Nicaragua*. `http://peacewinds.org/the-associativity-route-as-a-driver-of-development-study-on-cooperatives-in-the-central-northern-area-of-nicaragua/`, January 2011. Online; accessed 2016-04-13.

[9] Carol Schumacher, Martha Siegel, and Paul Zorn. *2015 CUPM Curriculum Guide to Majors in the Mathematical Sciences*. Mathematical Association of America, 2015.

[10] George B. Thomas, Maurice D. Weir, and Joel R. Haas. *Thomas' Calculus: Early Transcendentals*. Pearson, 13th edition, October 2013.

[11] Tracy Bibelnieks, Mark Lester, John Zobitz. *Math Students Go into the World: Internationalizing the Mathematics Curriculum with the Calculus of Sustainability*. MAA Focus, October/November 2012, pp 14-16.

Postscript

In this volume we have gathered together thoughtful essays on the role that mathematics instructors can play in the classroom to bring social justice issues to the fore and featured an eclectic collection of contributed materials which can be adapted to a wide range of mathematics courses at the collegiate level.

The term "math for social justice" has been more prevalent in the K-12 context, and interested readers may follow up with some of the K-12 resources we indicated in the introduction (Chapter 1). However we also hope that those who find the ideas presented here intriguing but not quite what they need for their own contexts will embrace this challenge head on and develop their own materials for their own students and institutions. In particular almost all the modules in this volume engage with current issues and events, and as such, the data they use, and sometimes the concerns they reflect, will likely invite updating over the years. This volume is not meant to be a timeless text; rather it is meant to inspire readers, to convince them that they too can be creative and resourceful as they bring into their own classes contemporary issues that most matter to them and their students.

When mathematicians talk about social issues in mathematics classes, they often think of general education or quantitative literacy courses, as well as courses in statistics and data analysis. It is true that these are natural contexts for engaging students mathematically with social justice issues. In fact the second volume of this project, *Mathematics for Social Justice: Focusing on Quantitative Reasoning and Statistics*, will include seventeen modules that can be used in just these kinds of courses. To provide for the reader a sense of what is in that book, we include below two tables from the introduction to that volume, which are analogues of Tables 1.1-1.2.

In this first volume, however, we hoped to show that social issues do not need to be limited to such courses. Resourceful instructors have been able to work with a much wider range of courses where these issues may be approached, addressed, and sometimes resolved. We see this as a hopeful trend, that mathematicians are taking active responsibility to engage with social issues on their own turf.

We are delighted to have a range of courses represented in this volume, but we are also quite aware that some undergraduate mathematics courses are visibly under- or un-represented. Consider, for instance, introduction-to-proofs-type classes. What kinds of classroom or out-of-class experiences can we devise that allow the students to connect their learning to the types of social issues they are keen on resolving? What kinds of projects may instructors assign that will allow students of number theory, say, to consider the social justice implications of their mathematical work? We trust that the mathematical community will have many creative answers to such questions.

– Gizem Karaali and Lily S. Khadjavi

Table P.1: Modules in *Mathematics for Social Justice: Focusing on Quantitative Reasoning and Statistics*, categorized by relevant mathematical courses.

COLLEGE ALGEBRA

"'I Need a Job!'": Analyzing Unemployment Rates in College Algebra and Introductory Statistics" by A. Brisbin (Chapter 2)

GENERAL EDUCATION / LIBERAL ARTS

"A Gentrification Module for Quantitative Reasoning" by F. Fisher and J. Warner (Chapter 4)
"Measures of Income Inequality" by A.J. Miller (Chapter 9)

"Super Size Me: Exploring the Nutrition of Fast Food" by K. Piatek-Jimenez (Chapter 10)
"Exploring the Benefits of Recycling" by K. Piatek-Jimenez (Chapter 11)
"The New Jim Crow: A Racial Checkup for the United States" by V. Piercey (Chapter 12)
"Who Makes the Minimum Wage?" by K. Simic-Muller (Chapter 13)
"Partisan Politics and Central Tendencies" by U. Williams (Chapter 17)

QUANTITATIVE REASONING

"Understanding Over- and Underrepresentation via Conditional Probability" by J. Belock (Chapter 1)
"A Three-Part Module on Poverty" by T.M. Brown (Chapter 3)

"A Gentrification Module for Quantitative Reasoning" by F. Fisher and J. Warner (Chapter 4)
"Examining Human Rights Issues Through the Lens of Statistics" by M. Franco (Chapter 5)
"Normal isn't 'Normal' When it Comes to Income" by T. Galanthay and T.J. Pfaff (Chapter 6)
"Traffic Stops and The Issue of Racial Profiling" by D. Greenberg, D. Hughes Hallett, and L.S. Khadjavi (Chapter 8)
"Measures of Income Inequality" by A.J. Miller (Chapter 9)

"Super Size Me: Exploring the Nutrition of Fast Food" by K. Piatek-Jimenez (Chapter 10)
"Exploring the Benefits of Recycling" by K. Piatek-Jimenez (Chapter 11)
"The New Jim Crow: A Racial Checkup for the United States" by V. Piercey (Chapter 12)
"Who Makes the Minimum Wage?" by K. Simic-Muller (Chapter 13)
"The Limits of Partisan Gerrymandering" by J. Suzuki (Chapter 15)

"Forecasting the Past: Teaching Regression" by Z. Teymuroglu and J.C. Chambliss (Chapter 16)
"Partisan Politics and Central Tendencies" by U. Williams (Chapter 17)

INTRODUCTORY STATISTICS

"'I Need a Job!'": Analyzing Unemployment Rates in College Algebra and Introductory Statistics" by A. Brisbin (Chapter 2)

"Examining Human Rights Issues Through the Lens of Statistics" by M. Franco (Chapter 5)

"Normal isn't 'Normal' When it Comes to Income" by T. Galanthay and T.J. Pfaff (Chapter 6)

"Get the Lead Out: The Connection Between Lead and Crime" by T. Galanthay and T.J. Pfaff (Chapter 7)

"Traffic Stops and The Issue of Racial Profiling" by D. Greenberg, D. Hughes Hallett, and L.S. Khadjavi (Chapter 8)

"The Limits of Partisan Gerrymandering" by J. Suzuki (Chapter 15)

"Forecasting the Past: Teaching Regression" by Z. Teymuroglu and J.C. Chambliss (Chapter 16)

"Partisan Politics and Central Tendencies" by U. Williams (Chapter 17)

PROBABILITY & STATISTICS

"Understanding Over- and Underrepresentation via Conditional Probability" by J. Belock (Chapter 1)

"Examining Human Rights Issues Through the Lens of Statistics" by M. Franco (Chapter 5)

"Normal isn't 'Normal' When it Comes to Income" by T. Galanthay and T.J. Pfaff (Chapter 6)

"Get the Lead Out: The Connection Between Lead and Crime" by T. Galanthay and T.J. Pfaff (Chapter 7)

"Traffic Stops and The Issue of Racial Profiling" by D. Greenberg, D. Hughes Hallett, and L.S. Khadjavi (Chapter 8)

"Should We Institute Mandatory Drug Tests for Recipients of Public Assistance?" by J. Suzuki (Chapter 14)

"The Limits of Partisan Gerrymandering" by J. Suzuki (Chapter 15)

DISCRETE MATHEMATICS

"Measures of Income Inequality" by A.J. Miller (Chapter 9)

CONTENT COURSES FOR TEACHERS

"Who Makes the Minimum Wage?" by K. Simic-Muller (Chapter 13)

Table P.2: Modules in *Mathematics for Social Justice: Focusing on Quantitative Reasoning and Statistics*, categorized by social justice theme clusters.

ACCESS
"A Three-Part Module on Poverty" by T.M. Brown (Chapter 3) (Chapter 3)
"A Gentrification Module for Quantitative Reasoning" by F. Fisher and J. Warner (Chapter 4)
"Normal isn't 'Normal' When it Comes to Income" by T. Galanthay and T.J. Pfaff (Chapter 6)
"Get the Lead Out: The Connection Between Lead and Crime" by T. Galanthay and T.J. Pfaff (Chapter 7)
"Super Size Me: Exploring the Nutrition of Fast Food" by K. Piatek-Jimenez (Chapter 10)

CITIZENSHIP
"The Limits of Partisan Gerrymandering" by J. Suzuki (Chapter 15)
"Partisan Politics and Central Tendencies" by U. Williams (Chapter 17)

(GLOBAL) CITIZENSHIP
"Exploring the Benefits of Recycling" by K. Piatek-Jimenez (Chapter 11)

ENVIRONMENTAL JUSTICE
"Get the Lead Out: The Connection Between Lead and Crime" by T. Galanthay and T.J. Pfaff (Chapter 7)
"Exploring the Benefits of Recycling" by K. Piatek-Jimenez (Chapter 11)

EQUITY / INEQUITY
"A Three-Part Module on Poverty" by T.M. Brown (Chapter 3) (Chapter 3)
"Normal isn't 'Normal' When it Comes to Income" by T. Galanthay and T.J. Pfaff (Chapter 6)
"Measures of Income Inequality" by A.J. Miller (Chapter 9)

FINANCE
"A Three-Part Module on Poverty" by T.M. Brown (Chapter 3) (Chapter 3)
"A Gentrification Module for Quantitative Reasoning" by F. Fisher and J. Warner (Chapter 4)
"Normal isn't 'Normal' When it Comes to Income" by T. Galanthay and T.J. Pfaff (Chapter 6)
"Measures of Income Inequality" by A.J. Miller (Chapter 9)
"Super Size Me: Exploring the Nutrition of Fast Food" by K. Piatek-Jimenez (Chapter 10)
"Who Makes the Minimum Wage?" by K. Simic-Muller (Chapter 13)